Let's talk an OIL DEAL!

Your key to oil patch lingo

by John Orban III

Meridian Press Oklahoma City

Published by
Meridian Press, Oklahoma City

First Edition - November 15, 1991

ISBN 0-9615776-2-2

This publication is designed to provide accurate and authoritative information in regard to the subject matter covered. It is sold with the understanding that this publication does not purport to render legal, accounting, or other professional service. If expert advice or assistance is required, the services of a competent professional should be sought.

A MESSAGE FROM THE PUBLISHER

Much of the material in this book has been high-graded from the glossary in *MONEY IN THE GROUND,* by John Orban, III

The idea for this printing came from David W. Kennedy, Executive Editor of the *LIMITED PARTNERSHIP INVESTMENT REVIEW*.

In reviewing the Second Edition of *MONEY IN THE GROUND*, he wrote "... there is also, by the way, an excellent glossary of oil and gas terminology, which is difficult if not impossible to find elsewhere."

A number of other reviewers also singled out the glossary *MONEY IN THE GROUND*, including Elizabeth Pike in the "Odd Lots" column of *REGISTERED REPRESENTATIVE* magazine, and Robin Robinson, in the "Oil Patch" column of *THE JOURNAL RECORD* (Oklahoma City).

This small stand-alone volume is designed to be pocket-handy ... and *affordable* ... for anyone with an interest in oil and gas.

MERIDIAN PRESS

Here is the language of oil . . .

Not complete by any means. That might require an entire section of a library. But this one book will help you understand the oil business.

With the exception of food and water, oil is the most important commodity in the world. You really *can't* leave home without it!

But until today, oil terminology has been a foreign language to the very people who perpetuate the industry . . . you and your family . . . and everyone else who uses and relies on petroleum products every day.

Now, in this handy guide, you hold the key to understanding oil and gas deals.

Whether you're ...a consumer ...a land-owner ...an investor ... or a professional in the industry, you'll be able to speak with authority, about the *business* side of the oil and gas business.

Let's talk an OIL DEAL!

Italicized terms are defined elsewhere in this book.

Please refer to *MONEY IN THE GROUND* for illustrations, diagrams, and examples of any item indicated by an asterisk (*).

ABANDON - A well is permanently *plugged* and abandoned if it is drilled and found to be a dry hole, or in the case of a producing well, if it ceases to be economically productive.

ABANDONMENT LOSS COSTS - *Lease* acquisition costs on *leases* that are abandoned, and non-salvageable equipment costs on dry holes, which provide *deductions* against *gross income*. In oil and gas partnerships, they are generally allocated between the *general partner* and *limited partners* according to who paid the costs.

ABSTRACT OF TITLE - A chronological history of the ownership of a tract of land. It may consist of all recorded legal documents affecting the title of ownership of a tract of land, from one owner to another, dating back to when the land was originally home-steaded or granted by the government, or it may be abbreviated to some degree. (Generally, 'recorded' means recorded at the courthouse of the county in which the land is located).

ACIDIZING - Pumping acid (usually hydrochloric acid) into a *reservoir*. As the acid dissolves *calcite*, the naturally occurring holes in the rock are are opened and enlarged, facilitating increased flow from the *reservoir*. *Limestones* are frequently treated with acid. *Sandstones* may be treated with acid if they contain calcite. (The acid does not affect the quartz grains making up the *sandstone*.) See also *fracturing*.

ACRE - The most common unit of land measure in the United States. A square 210 feet on a side (44,100 sq ft) would be a bit larger than an *acre* (43,560 sq ft). There are 640 acres in a square mile. *

ACRE-FOOT - In the U.S., the thickness of a *pay zone* is measured in feet, and the area of the *reservoir* is measured in *acres*. An acre-foot is a volume of *reservoir* rock that is one *acre* in area and one foot thick. Estimates of recoverable oil or gas from a particular *reservoir* are generally expressed in *barrels* of oil per acre-foot, or in *MCF* (thousands of cubic feet of gas) per acre-foot. Commercial oil recoveries typically range from about 50 to 500 barrels per acre-foot, gas recoveries range from about 10 to 1000 MCF (thousands of cubic feet) per acre-foot. *

ADJUSTED BASIS - The *basis* (or cost) of property increased (for *capital improvements*) or decreased (for investment *tax credits* and *depreciation allowance* adjustments to gross income) used in computing

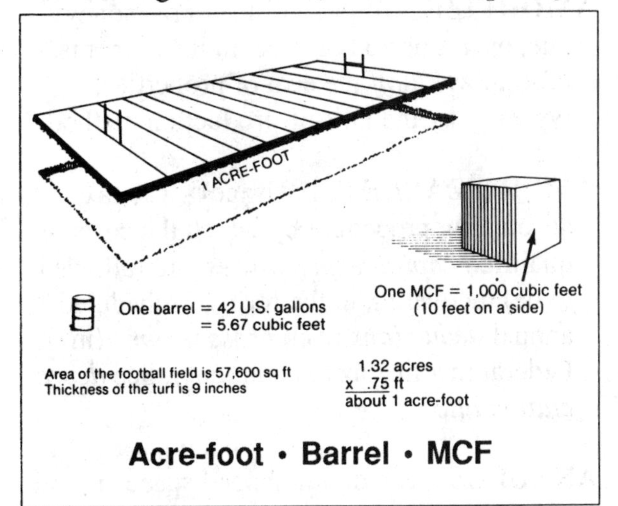

Acre-foot • Barrel • MCF

gain or loss, for income tax purposes. See *basis*.

AFE - See *Authorization For Expenditure*.

ALLOWABLE - The maximum amount of oil or gas a well or field is permitted to produce per day. It is typically set by state regulating agencies, after considering the economic market for the product, the *maximum efficient rate* (MER), and other factors.

ALTERNATIVE MINIMUM TAX - A tax imposed when an individual's *tax preference* items exceed: his regular tax plus his regular minimum tax liability.

AMORTIZE - To pay off a sum of money due, over a period of time, in installments. Also to *write-off* the cost of properties, typically by the unit-of-production method.

AMORTIZABLE (Tax Usage) - The tax accounting procedure by which the costs of qualified *capital expenditures*, are (allowed to be) recovered, in the form of scheduled annual *deductions* from *gross income* (in Federal income tax calculations). See *depreciation allowance*.

ANNULUS - The donut-shaped space around

the outside of the *drill pipe* in an open *borehole*. A *cased well* has two annular spaces: one between the *drill pipe* and the *casing*, the other between the *casing* and the *borehole*. *

ANTICLINE - The 'textbook' type of (structural) oil or gas *trap*. It consists of layers of rock that have been *folded* into a domal shape, that would look like an inverted soup plate. Oil or gas can be trapped in an anticline, the way air might be trapped inside an over-turned rowboat.

API - American Petroleum Institute, a petroleum industry association that sets standards for oil field equipment and operations. See *oil gravity*.

ARTIFICIAL LIFT - The use of mechanically applied power to lift oil to the surface in a producing well.

ASSESSMENTS - Additional capital *contributions* that may be become required of the *limited partners* during the course of the partnership's existence, after capital formation of a *limited partnership*. The amount and circumstances (voluntary, mandatory, etc), relating to assessments are addressed in the partnership agreement. Although an assessment may be called 'voluntary' or 'optional', the *limited partnership* agreement may state that the investor will suffer a penalty if he does not pay it.

ASSET - Any physical property or legal right that is owned, can be legally transferred to another party, and has a monetary value.

ASSIGNMENT (ASSIGN) - Legal document transferring an interest in a property from one party to another. The receiving party is the 'assignee'; the transferring party is the 'assignor'.

ASSOCIATED GAS - *Natural gas* that naturally occurs in a *reservoir* along with ('associated' with) *oil*, in a *gas cap*, or dissolved in the oil.

ASSOCIATION FOR TAX PURPOSES - When two or more taxpayers are engaged (associated) in a common undertaking, if their association resembles a corporation (even though it may not be one), the association might be taxed as a corporation.

AT-RISK LIMITATION - The amount of (tax) losses *deductible* from *gross income* is limited to the amount the taxpayer has invested (in cash, property, or recourse debt for which he is liable) that is at *risk* of being lost.

AUTHORIZATION FOR EXPENDITURE (AFE) - An estimate of the costs of drilling and completing a proposed well, which the operator provides to each *working-interest* owner before the well is commenced. Various categories of costs are typically listed as 'dry hole' costs (the costs to drill to the *casing* point; these are costs that would be incurred if no indications of *hydrocarbons* are found), *completion* costs (the additional costs to complete the well), and the total cost. *

BACK-IN - A type of interest in a well or property that becomes effective at a specified time in the future, or on the occurrence of a specified future event. See *reversionary interest*. *

Please refer to MONEY IN THE GROUND for more detailed discussion and examples of various back-in deals.

BARREL (BBL) - The standard unit of oil measurement in the U.S. oil industry. A barrel equals 42 U.S. gallons, or 159 liters. [In some countries oil is measured using the metric system: volumes in liters, or weights in metric tons.] *

BASEMENT - (Usually) non-sedimentary rock in which oil is not likely to be found. It underlies the various layers of *sedimentary rock*.

BASIN - A natural depression on the earth's surface, in which (usually water-borne) sediments accumulate over millions of years. The Gulf of Mexico is an example. Several hundred sedimentary basins have been identified around the world. *

Please refer to MONEY IN THE GROUND for map and listing of oil/gas-producing basins in the U.S.

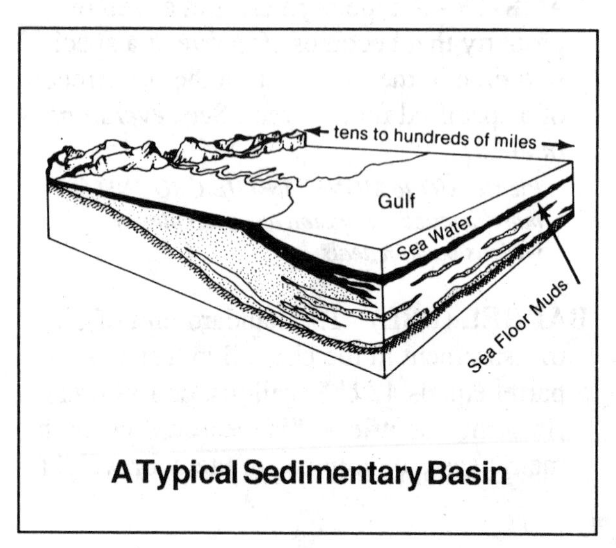

A Typical Sedimentary Basin

BASIS - The cost of property that is used in computing gain or loss, for Federal income tax purposes.

BCF - Billions of cubic feet (of gas). In the U.S. oil and gas industry, the cubic foot is the standard unit of measurement of gas at atmospheric pressure. Billion is 1,000 million (10^9).

BED - A layer of rock.

BEHIND PIPE - If a well drills through several *pay zones* and is *completed* in the deepest productive *reservoir*, *casing* is set all the way down to the producing *zone*. Viewed from (a perspective) inside the *borehole*, *reserves* in the shallower *pay zones* up the hole are **behind** the *casing* **(pipe)**. *

BLIND POOL - Refers to an oil and gas *limited partnership* which has not committed to specific *prospects*, *leases*, or properties at the time of capital formation (*subscription* by the *limited partners*).

BLOWOUT - A sudden uncontrolled flow from a well into the atmosphere (of oil, gas, water, or mud). It can occur when a *borehole* is drilled into a rock-layer in which

Unidentified Blowout

Photo Courtesy of the
Oklahoma State Historical Society

natural pressures are greater than that exerted by the column of drilling *mud* in the borehole. The force of the uncontrolled flow from a blowout is dangerous enough; if gas is present, there is the additional danger of explosion and fire. *

BLOWOUT INSURANCE - An insurance policy that protects the insured party (*working-interest* owner) from liabilities which might arise from a *blowout* during the drilling, completion, or production of a well.

BLOWOUT PREVENTERS - Heavy-duty equipment installed at the wellhead during drilling and completion operations, to prevent the possibility of a *blowout*, by sealing the *annular* space between the *drill pipe* and the *casing*. Pipe rams, shear rams, or blind rams may be activated to choke off flow from the *borehole*.

BLUE SKY LAWS - State regulations governing an offering to sell *securities* within that state (analogous to Securities Exchange Commission registration requirements at the Federal level). The expression originated with the first state securities laws enacted in the U.S. (Kansas, in 1911), intended to protect state residents from being sold *securities* as lacking in substance, 'as the blue sky'.

BOILER ROOM - Nowadays it might be called a 'tele-marketing' operation. The term refers to a phone room manned by sales pros, who call prospective investors and try to interest them in oil and gas deals. Since *securities* regulations place restrictions on the distribution of offering documents (especially in the case of a *private placement*), the initial approach is often made by phone.

BONUS - Cash paid to the *mineral-rights* owner (lessor) for an oil and gas *lease*. The bonus is a negotiable provision of the *lease*. See *lease*.

BOREHOLE - The hole created by the drilling (boring) of a well. *

BS&W - Bottom sediment and water that accumulates in the bottom of an oil tank (see illustration, page 24). More generally anything that has no value and is a nuisance to dispose of.

BOTTOM-HOLE - An adjective referring to the deepest part of a borehole.

BOTTOM-HOLE PRESSURE - The pressure in a well, measured by an instrument that is lowered into the bore hole on a *wireline*. It may be measured under flowing or *shut-in* conditions.

BOTTOMS-UP - A temporary interruption in drilling (several hours in a deep well) to allow *cutting* samples to be *circulated* from the bottom of the hole up to the surface. The purpose is to allow the geologist at the well-site to inspect the cutting samples before giving permission to resume drilling. If the cuttings contain *shows* of oil or gas, special procedures such as *coring* or *drill stem testing* may be desirable before drilling is resumed. See *drilling break*.

Btu - British thermal unit, a unit of heat energy, used to describe the amount of heat that can be generated by burning oil or gas. The heat generated from lighting a kitchen match is about 1 Btu; the heat needed to make a cup of coffee would be about 60 Btu.

CALCITE - The mineral calcium carbonate, $CaCO_3$. It is the primary ingredient of *limestone*, and is also present in dolomite, and often in *sandstone*.

CALIPER LOG - A *wireline log* that graphically indicates variations in the diameter of the borehole, displayed versus depth. See *logs*.

CAPITAL - Funds invested in a business for use in conducting the operations of the business.

CAPITAL ASSET - An *asset* acquired as an investment, for the purpose of creating a product or service intended to be used in the activities or operations of a business. (Not an *asset* intended for resale). A *royalty interest* held for investment would be a capital asset.

CAPITAL COSTS (Oil & Gas Tax Usage) - For Federal income tax purposes, the costs of *capital expenditures* which may be recovered by *deduction* against income (through *depreciation* and *depletion*).

CAPITAL EXPENDITURE - An expenditure intended to benefit the future activities of a business, usually by adding to the *assets* of a business, or by improving an existing *asset*. Such an expenditure is treated like an investment expected to generate future income for the business.

CAPITALIZE - To treat certain expenditures

as capital expenditures for Federal income tax computations.

CARRIED INTEREST - A fractional *working interest* in an oil and gas lease that comes about through an arrangement between co-owners of a *working interest*.

The owner of a carried interest is not liable for certain development and operating costs. The co-owner(s), who assumes these obligations and advances the cost of them, may or may not recoup his advances out of the carried party's share of production revenues (if any), according to the carrying arrangement.

In the case of a single well, the carry most typically includes all costs incurred in drilling to the *casing point*, but it could include any or all phases of the drilling, *completion*, and flowline hook-up of a well.

In the case of leased property, the carry can include costs incurred in any or all phases of exploration and development: from *prospect* generation and *lease* acquisition, through drilling the initial well on a *prospect*, through drilling a specified number of wells on a *prospect*, through the life of the *leased* property.

At the point of carry, the carried party's interest typically converts to a full *working interest*, although many other possibilities could be specified in a carrying arrangement. Details of the carried party's entitlement to revenue can vary considerably: starting from initial production, or not starting until the carrying party has recouped out of production, the costs he advanced, or some other arrangement.

Abercrombie-type, Herndon-type, and Monahan-type are types of carried-interest arrangements (named after parties involved in court decisions of the same name). They are discussed at length in Burke and Bowhay's "Income Taxation of Natural Resources".

Please refer to MONEY IN THE GROUND for more detailed discussion and examples of carried interest deals.

CARVED OUT INTEREST - A fractional interest transferred (conveyed) to another party by the original owner of the whole interest. See *retained interest.*

CASED HOLE - A *borehole* in which *casing* has been run. If the *casing* does not extend all the way to *total depth* (the bottom of the hole), the un-*cased* portion is *open hole.* *

CASH DISTRIBUTIONS - Moneys paid by an oil and gas partnership to its partners according to the terms of the partnership

agreement. These distributions are net of state *severance* tax, and *windfall profits tax*).
Up to the amount of a partner's tax *basis* in the partnership, such distributions are generally held to be a return of investment (hence, not income). They are not taxable to the recipient until their (cumulative) total exceeds the partner's tax basis in the partnership. Thereafter, the distributions represent true income, and the receiving partner must pay income tax on distributions received.

CASING - Large diameter steel pipe placed in a *borehole* to support the sides or walls of the hole and to prevent them from caving in. It not only prevents the loss (flow) of drilling fluids from the *borehole* into rocks penetrated by the *borehole*, but also prevents rock fluids from flowing into the *borehole*.
Individual sections are usually about 30 feet long, and are screwed together as they are lowered (or 'run') into the hole. Once casing is run in the hole, it is *cemented* in place by pumping a slurry of *cement* into the space between the outside of the casing and the walls of the *borehole* (the *annulus*).
Casing run from the surface is called surface casing. Additional strings of 'intermediate casing' may be required during the drilling of the well. Production casing is run in order to *complete* a well. *

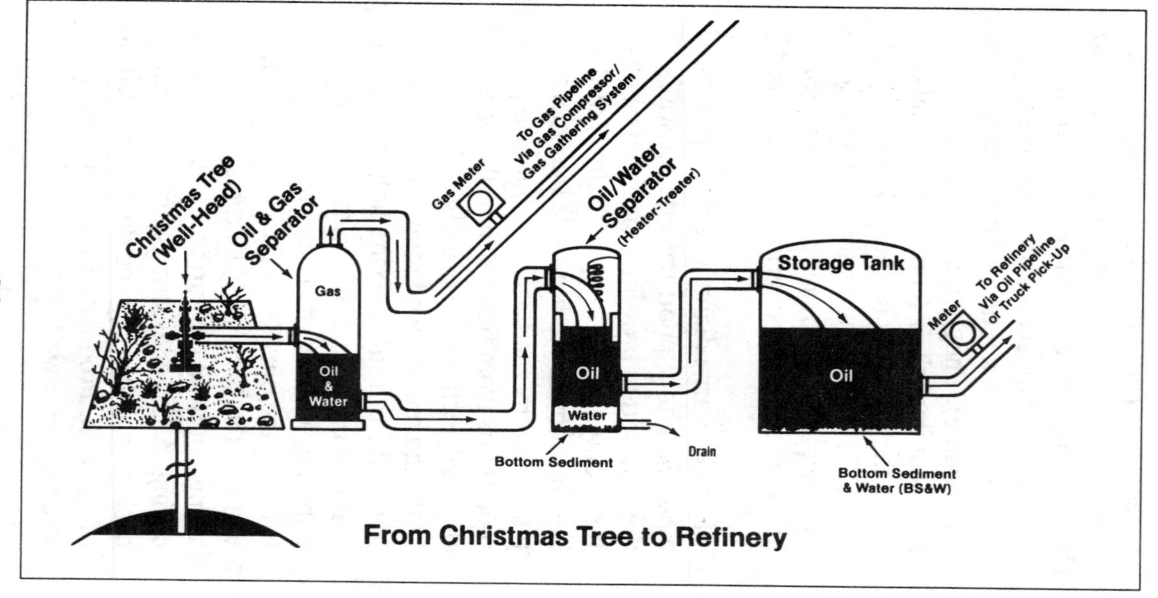

From Christmas Tree to Refinery

Please refer to MONEY IN THE GROUND for a complete discussion of the 'casing point election'.

CASINGHEAD - A fitting attached to the top ('head') of the *casing* in a producing oil or gas well (usually at or just above ground level). It regulates the flow of oil or gas, helps separate oil from gas, allows the pumping of oil from an oil well, and facilitates periodic cleaning out of the well. See *Christmas tree*.

CASINGHEAD GAS - *Natural gas* produced along with oil from an oil well (as distinguished from gas produced from a gas well). CASINGHEAD OIL is oil produced from a gas well.

CAVINGS - Rock fragments that break off from the walls of a borehole and fall into the borehole during drilling operations.

CEMENT - Fluid cement is mixed at the surface, pumped to the bottom of a *cased* well, forced to flow around the lower end of the *casing* and up into the space between the *casing* and the *borehole*. When the cement solidifies (sets), it holds the casing in place, and provides support.

CHRISTMAS TREE - A complex assemblage of valves/controls, fittings, gauges, and pipe connections at the top of the *casing* of a flowing oil well. It controls the flow of oil and gas from the well. A silhouette of this assemblage might be vaguely reminiscent of a heavily decorated Christmas tree, hence the name. *

CIRCULATION - The continuous pumping of drilling fluid ('*mud*') from *mud* tanks at the surface: down through the *drill pipe*, out the nozzles in the *drill bit*, and back up to the surface through the space between the *drill pipe* and the *borehole*. The flow of *mud* moves the rock *cuttings* (and *cavings*) away from the *drill bit*, and carries them up to the surface, where they are strained out of the *mud* system, by the *shale shaker*.

'Lost circulation' occurs when drilling fluid escapes from an un-*cased* borehole into porous zones, or holes such as fractures or caverns that occur naturally in the penetrated rocks. Various plugging or clogging materials may be added to the *mud* in hope of stopping its loss. (As one might inject gummy fluid into an automobile tire to stop a leak.) Difficult (and often costly) mechanical operations may be required to drill through a *zone* of lost circulation. If an attempt to drill through zones of lost circulation are unsuccessful, a well might have to be *abandoned.**

CNG - Compressed Natural Gas.

COGENERATION - A process of obtaining more efficient use of fuel. The energy produced is used not only to generate electric or mechanical energy, but also to provide useful thermal energy (such as heat and steam), for use in industrial, commercial, heating or cooling processes.

A topping cycle is when the fuel is first used to generate electric or mechanical power, and the by-product (reject heat) is then used to provide thermal energy.

A bottoming cycle is when the fuel is first used to provide useful thermal energy, and the by-product (reject heat) is then used to generate electric or mechanical power.

COMMISSIONS - Payments to qualified agents of the sponsor of a *limited partnership*, for selling interests in it to investors. The agents typically include officers and directors of the sponsor, underwriters, securities dealers, and other qualified third parties. Commissions may take the form of a percent of partnership interests sold, an oil and gas interest, or stock in the sponsor's company. The costs of the commissions are not *deductible expenses*; they are non-*amortizable capital expenditures*.

COMMON LAW - Social customs and behavior patterns established in the past, that come to be regarded and accepted as law (on the basis of court decisions and precedent, rather than legislative action).

COMPLETION (COMPLETE A WELL) - After the drilling of a successful well, the 'completion' includes all the work required to make the well ready for commercial production.

CONDENSATE - *Hydrocarbons* naturally occurring in the gaseous phase in the *reservoir* that condense to become a liquid at the surface (due to the change in pressure and temperature).

CONTRIBUTION - In oil and gas *limited partnerships*: the payment of money, property, or services by which an investor becomes a *limited partner* in the partnership.

CONVERTIBLE INTEREST - An interest (usually a non-cost-bearing interest) that may, at the option of the owner or on the occurrence of a specified event, be changed into another type of interest (usually a cost-bearing interest). Example: a 5% *overriding royalty* convertible to a 1/8 (= 12.5%) *working interest* after *payout.* *

Please refer to MONEY IN THE GROUND for examples of convertible interests.

CONVEY (CONVEYANCE) - Legal term for transferring the title of a property from one party to another, typically by *deed.*

CORE - A cylindrical column of rock usually 4 to 6 inches in diameter cut in lengths of about 30 feet by a special *drill bit.* (The operation is a bit like removing the core from an apple). After the core has been brought to the surface, it is examined by the geologist for *shows* of *hydrocarbons.* Because of the time and high costs cost involved, only the most critical intervals are cored in the U.S. (In China, however, it is not uncommon to core an exploratory well from grass-roots to granite.)

CORRELATIVE RIGHTS - A *mineral-rights* owner is entitled to capture oil and gas from an accumulation which exists beneath his property, but extends beyond his property line. He has the ('correlative') right to drill a well on his property, and produce oil and gas which might otherwise flow toward wells (drilled on adjacent land owned by others), that produce from the same oil and gas accumulation.

"CREEK-OLOGY" - Refers to locating the next well, not strictly on the basis of science and geology, but also in part, on the basis of gross natural features on the surface of the land. (The bend in a creek, for example, may represent the surface expression of a fault or a structural trap deep beneath the surface). A large corporation might never admit it to its shareholders, but even after all the scientific data are collected and analyzed (at considerable expense) the results are often not conclusive; the final decision on where to drill may include a good measure of "creek-ology".

CRUDE OIL - A naturally occurring mixture of liquid *hydrocarbons* as it comes out of the ground (before or after any dissolved gas has been separated from it, but prior to any process of distilling or refining). Greenish crude is usually high in paraffin (wax) content; blackish oil is more likely to be asphaltic. Different types of *source rock* generate different types of crude oils.

CURRENT - refers to the present (tax or fiscal) year.

CUTTINGS - Small chips of rock (usually about the size of a fingernail) produced by the grinding motion of the *drill bit* during drilling. The drilling *mud* carries ('*circulates*') them to the surface where they are strained out of the *mud* system by the *shale shaker*. Cuttings are examined under a microscope for direct observation of *porosity*. When placed under ultraviolet light, any *hydrocarbons* contained in the pore spaces of the cuttings will give off a distinctive glow ('*fluorescence*'). This fluorescence is recorded as a *show* of oil.

DEDUCTIONS - Tax items which may be subtracted from *gross income* to arrive at *taxable income* in Federal income tax computations. See *taxes due*.

DEED - A written document (legal 'instrument') by which the title to a property is transferred (*conveyed*) from one party (the grantor) to another (the grantee). It is signed by the grantor and delivered to the grantee.

DELAY RENTAL - Periodic cash payments to the *mineral-rights* owner (lessor), by the *working-interest* owner (lessee), for the privilege of postponing the commencement of drilling operations on the *leased* property. (Delay rentals are usually specified as an annual payment per *acre* of land covered by the *lease*). A 'paid up *lease*' means that all the delay rentals were paid up-front. Delay rentals are a negotiable provision of an oil and gas *lease*.

DEPLETION - The value of a naturally occurring mineral deposit is a function of (1) the market value of the mineral, and (2) the concentration of the mineral in the mineral deposit. Physical depletion is the exhaustion of a mineral deposit through production of the mineral. Economic depletion is the reduction in the value of the mineral deposit as it becomes exhausted (less concentrated) through production.

DEPLETION ALLOWANCE - The income tax *deduction* allowed for the exhaustion of a natural resource. *

DEPRECIATION - A tax accounting method in which the value of an *asset* (starting with its acquisition cost) is reduced each year. Scheduled annual (non-cash) amounts are charged against the asset, representing the gradual loss of value, as it wears out, deteriorates, or becomes obsolete. (Theoretically such charged amounts accumulate over time, providing funds that could be used to replace the worn out *asset*.)

DEPRECIATION ALLOWANCE - In the computation of taxable income: a *deduction* from *gross income* allowed during the useful life of an *asset* (in recognition of its wear and tear, and eventual obsolescence). Different types of property (*assets*) have different years of useful life. This determines the methods (schedules) for calculating the deduction (straight line, declining balance, sum-of-the years-digits, etc).

DIP - The angle that a rock layer happens to be inclined (at a specific location), measured relative to a horizontal plane (in degrees).

DIPMETER LOG - A wireline log that tells the angle and direction of the *dip* of rock-layers penetrated by the *borehole*.

DIRECTIONAL DRILLING - When a drilling rig cannot be positioned directly over a prospect, because of natural or man-made obstacles (buildings, a park, a swamp, a lake, etc.), it may be positioned off to one side. A *borehole* can be drilled at an angle (a 'slant hole') to reach the subsurface objective. See *whipstock*.

DIVIDED INTEREST - See *undivided interest*.

DIVISION ORDER - A contract for the sale of oil or gas, by the holder of a revenue interest (*royalty*, *overriding royalty*, *working interest*, etc.) in a well or property, to the purchaser (often a pipeline transmission company). It lists the names of revenue interest owners of a producible oil or gas well, along with their respective shares of production revenues, and directs the purchaser to distribute the proceeds of production sales, accordingly. *

Please refer to MONEY IN THE GROUND for further discussion of division orders.

DOE - U.S. (Federal) Department of Energy.

DOODLEBUGGER - A *geophysicist* who interprets *seismic data*. 'Doodlebug' originally referred to any device (such as a divining rod or forked stick used to search for water) applied to the search for oil. Although it came to include any new method used in the direct detection of oil and gas accumulations, now it generally refers to the field of *seismic* interpretation.

"DOODLEBUG-OLOGY" - Can refer to any new and un-proven method or technology (electrical, chemical, magnetic, etc.) used to search for oil and gas accumulations.

DOWNHOLE - Refers to equipment or mechanical operations that take place, **down** inside a bore**hole** (as contrasted with those at the surface).

DOWN TIME - **Time** lost during drilling, usually as a result of equipment break**down**.

DRILL BIT - A tool with very tough steel or diamond teeth that grind rock into small chips during drilling. The diameter of the bits used to drill a well may range from more than 22 inches at the upper part of the hole, to less than than four inches at *total depth*. *

DRILL PIPE - A special grade of extra-strong steel pipe threaded on both ends that comes in lengths of about 30 feet and diameters from about 6 1/2 to 2 1/2 inches. See *drill string*.

DRILL-STEM TEST (DST) - When a well is drilled into a potential *pay zone*, it may be desirable to try to make the zone flow, before continuing to drill ahead. To 'test' the zone, drilling is halted, and the string of *drill pipe* is pulled out of the *borehole*. The *drill bit* is removed and a special testing device is attached to the end of the *drill string* which is lowered back into the *borehole*.

A system of blocking devices (*packers*) allow rock fluids to flow from the reservoir directly into the drill pipe (drill stem). The quality and amount of formation fluids recovered, along with the pressures recorded during the drill stem test indicate whether the *zone* could be commercially produced. (An alternative to drill stem testing is to continue drilling to the planned *total depth*, and then *log* the hole. If the *logs* suggest a good *zone*, *casing* may be set. Then the *zone* can be tested through *perforations* made in the *casing*. Although this is often a more conservative procedure than taking a drill-stem test during drilling, it can be much more expensive.)

DRILL STRING - The entire assemblage of joints of *drill pipe* that are screwed together and lowered into a *borehole*. The drill string runs from the *Kelly* on the *drilling* rig to the *drill* bit at the bottom of the *borehole*.

DRILLING BREAK - A sudden increase in the rate of drilling. Usually it indicates that the *drill bit* is penetrating 'weaker' rock. The 'weakness' of the rock is presumed to result from increased *porosity* (which might contain hydrocarbons). See *bottoms up*.

DRILLING RIG - The surface equipment

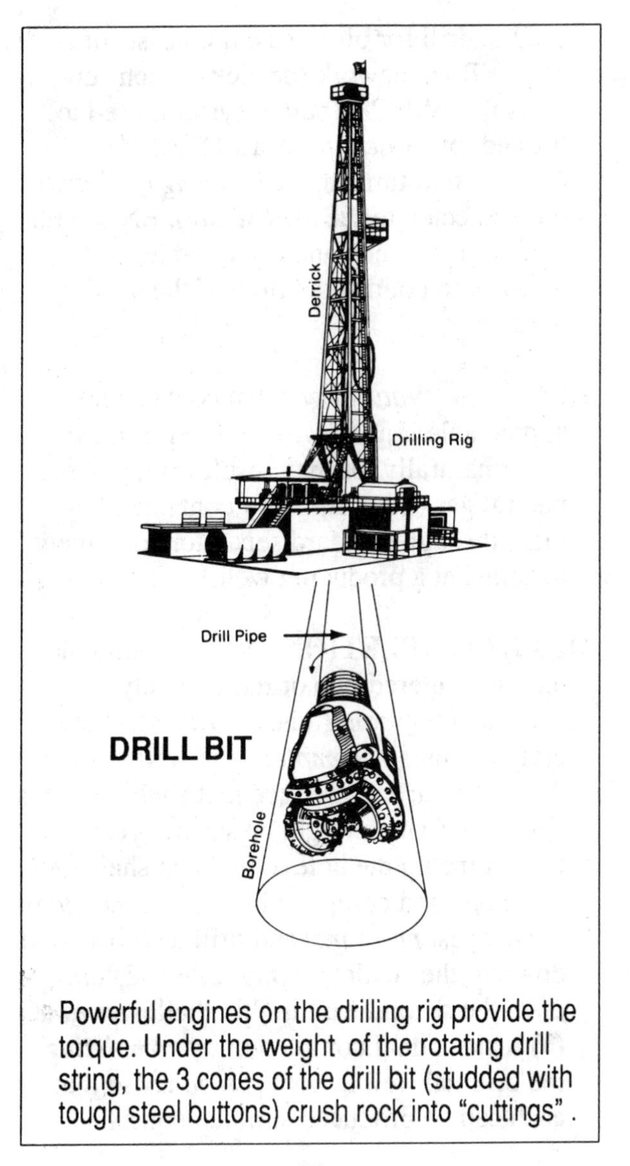

Powerful engines on the drilling rig provide the torque. Under the weight of the rotating drill string, the 3 cones of the drill bit (studded with tough steel buttons) crush rock into "cuttings".

used to drill for oil or gas. It consists of a TOWER framework (derrick) which supports the WINCH (pulley system) used to lift and lower *drill pipe*, a ROTARY TABLE that turns the *drill string* (and *drill bit* connected to the *drill string*), ENGINES to drive the winch and rotary table, and PUMPS to control the flow of the drilling *mud*. *

DRY GAS - *Natural gas* that contains no appreciable liquid *hydrocarbons*. It may occur naturally, or may result from passing natural gas (that originally contained liquids) through standard separator equipment installed at a producing well.

DUAL COMPLETION - When a borehole has encountered two or more widely separated *pay zones*, the *operator* has several options: (1) *complete* the well in the deepest *zone* and produce that *zone* until it is *depleted* (which might be several years), then come up the hole to the next shallower *pay zone* and complete in it; (2) *complete* in the deepest *pay zone* and drill a second well down to the shallower *pay zone* and *complete* the second well in this shallower *zone*; (3) make a dual completion by *completing* the original well in both *pay zones* so that each can be produced simultaneously through separate strings of *tubing*.

DUE DILIGENCE - In an offering of *securities*, certain parties who are responsible for the accuracy of the offering document, have an obligation to perform a 'due diligence' examination of the issuer: issuer's counsel, underwriter of the security, brokerage firm handling the sale of the security. Due diligence refers to the degree of prudence that might properly be expected from a reasonable man, on the basis of the significant facts which relate to a specific case (in other words, not measured by any absolute standard).

ECONOMIC INTEREST - An interest in oil and gas in the ground. It entitles the owner to a *deduction* from *gross income* derived from production of that oil or gas (*depletion allowance*). As specified in Federal income tax regulations, "an economic interest is possessed in every case in which the taxpayer has acquired by investment any interest in minerals in place or standing timber and secures, by any form of relationship, income derived from the extraction of the mineral or severance of the timber, to which he must look for a return of his capital."

ENHANCED OIL RECOVERY (EOR) - Various methods of increasing the recovery of oil and gas from a *reservoir*, (usually) after *primary recovery* has pretty much run its course. In the case of certain (low-pressure) *reservoirs*, or (highly *viscous*) oils, some of these techniques may be necessary, right from the beginning of production. See *flooding*, *gas injection*, *primary recovery*, *secondary recovery*.

EOR - *Enhanced oil recovery.*

ERTS - Earth Resource Technology Satellite. As commonly used, ERTS refers to satellite images of the earth's surface, digitally-recorded at various non-visible wavelengths (in the infra-red part of the spectrum).
Image enhancement by computer processing yields 'false-color images' that can reveal subtle features on the earth's surface which cannot be observed from the ground, or even on aerial photographs. Some of these features might be interpreted as the surface expression of geologic structures at depths of thousands of feet below the ground (which could contain oil or gas). See *trap*.

EUR - Expected (or estimated) Ultimate Reserves. An estimate of the cumulative volume of *reserves* that will ultimately be recovered (from a specified *reservoir*) over the life of a well, field, or property.

EXPENSES (Tax Usage) - Expenditures for business items that have no future life (such as rent, utilities, or wages) and are incurred in conducting normal business activities. In computing Federal income *tax due*, expenses may be deducted from *gross income* to arrive at *taxable income*. See *taxable income.*

FARM OUT AGREEMENT - An arrangement in which the responsibility of exploration and development is shifted (by assignment) from the working interest owner BOB, to another party AL. AL wants to drill a well on BOB's property. Under the farmout agreement: AL is obligated to perform a specified exploration and drilling program to earn a working interest in BOB's property; BOB retains an interest in the net proceeds of future oil or gas production from the property.

In this case, AL farmed into BOB's property; BOB farmed out his interest to AL. AL is the 'farmee'; BOB is the 'farmor'.

FAULT - A crack or fracture in the earth's

crust, along which the rocks on one side have moved relative to the rocks on the opposite side. The movement could be sideways or up and down, and can range from a few inches to tens (or even hundreds) of miles. A fault can juxtapose a non-porous layer of rock against an oil-bearing reservoir (sometimes thereby forming a *trap*).

FEE OWNERSHIP - The ownership of full right, *title*, and interest to the surface of a tract of land and to all minerals beneath it, as well as the air space above it. FEE SIMPLE OWNERSHIP (FEE SIMPLE ABSOLUTE) is fee ownership without any limitation or restriction as to future transfer of the ownership.

FEET OF PAY - The thickness of the *pay zone* penetrated in a well. In the case of an *oil column* floating on water, it is the thickness of the layer of oil ('the *oil column*') above the *oil-water contact*. *

FERC - Federal Energy Regulatory Commission.

FIELD (as in OIL FIELD or GAS FIELD) - A commercial oil or gas accumulation (or the land area above it). The size of an oil field can range from a few *acres*, to the

Ghawhar oil field in Saudi Arabia which is over 100 miles long and up to a dozen miles wide. *

FISHING - The procedure of locating and attempting to retrieve any object (a 'fish') that has accidentally fallen into, or been left in the *borehole*, and must be retrieved before mechanical operations (drilling, *logging*, *completion*, etc) can be resumed.

FLOODING - One of the methods of *enhanced oil recovery*. Water flooding or gas flooding might be considered *secondary recovery* methods; miscible flooding or chemical flooding, tertiary recovery methods.

The general procedure involves pumping (injecting) a fluid (commonly water) into the reservoir, through wells located around the perimeter of an oil field (peripheral drive). The 'pressure front' that is created, flushes oil toward the central part of the field, resulting in increased production. *

'FLOW THROUGH' CONCEPT - In ventures structured as partnerships (or S corporations), certain items of tax significance (profit, loss, etc.) are passed on to the partners (or S corporation shareholders) in the venture. In a venture structured as a 'C' corporation, the responsible tax-paying party would be the corporation itself (not its shareholders).

FLUORESCENCE - An optical property of some materials: they glow emitting visible light when they absorb radiation from an ultraviolet source. Liquid *crude oils* fluoresce with colors that range from brown to yellow to green to blue. The color may give some indication of the density of the oil and its chemical characteristics.

FOLDING - The bending of layers of rock, which occurs naturally, in response to pressures and stresses in the earth's crust. *

FORMATION - A layer of rock having characteristics that are distinct and recognizable. The rock layer is thus mappable, even among other layers of similar rocks. The thickness can range from a few feet to hundreds of feet. Distinctive features might include mineral composition, texture, diagnostic plant or animal remains (fossils) contained in it, etc.

FRACTURING - A procedure undertaken to attempt to increase the flow of oil or gas from a well. A fluid (usually crude oil, diesel oil, or water) is pumped into the *reservoir,* with such great force that the *reservoir* rock is physically broken and split open. Usually the 'frac fluid' carries small pellets or beads mixed in with it; the idea is for them to get caught in the fractures and prop them open (the beads or pellets are called the propping agent or *proppant*).

As the pumping pressures are gradually released at the surface, the natural *reservoir* pressures will force the 'frac fluid' out of the *reservoir*, and back into the well as the well begins to flow. The *proppant* remains behind, holding the fractures open, thereby increasing the flow of oil or gas from the reservoir into the well. This procedure is also called hydraulic fracturing. 'To frac a well' means to hydraulically fracture a reservoir in a well. *

FRONT-END COSTS - Costs that are paid out of initial investment in a venture, first, before the venture activities actually begin. In a *limited partnership*, these costs might include *syndication expenses*, legal fees, accounting fees, *management fees*, etc.

Front-end costs reduce the net amount available for investment (for the drilling of wells, purchase of producing properties, purchase of royalties, etc.)

FUTURE NET REVENUES - Net revenues

The Grand Canyon *Photo Coutesy of John Halsey*

FLAT-LYING ROCK LAYERS

Lulworth Cove, U.K. *Photo by Author*

FOLDED ROCK LAYERS

that will be received in the future. After discounting back to the present value of such revenues, the amount is often expressed as the "present value of future net revenues".

FUTURES PRICES - Refers to the New York Mercantile Exchange (NYMEX) which introduced futures contracts for *crude oil* in 1985 and *natural gas* in 1990. A futures contract is an obligation to buy or sell a specified quantity at a specified price in some future month (as far as 18 months in the future).

An oil futures contract is for 1,000 barrels of *West Texas Intermediate* (*WTI*) *crude oil* to be delivered at Cushing Oklahoma; price is quoted in dollars per barrel. WTI is the domestic light, sweet crude that is most widely traded in the *spot market*. See *API*, *sour crude*, *WTI*.

A gas futures contract is for 10,000 MM *Btu* of gas to be delivered during the calendar delivery month, at the "Henry Hub" near Erath, Louisiana; price is quoted in dollars per MM *Btu*. (MM= 1,000,000.)

GAS CAP - An accumulation of natural gas, above a layer of *saturated* liquid *hydrocarbons*, in a *reservoir*. *

GAS COLUMN - The vertical height of a gas accumulation above the gas-oil or gas-water contact. In commercial gas fields, a gas column can range from several feet to (rarely!) more than a thousand feet. See *oil column.*

GAS CONDENSATE - See *condensate.*

GAS DRIVE - When a well drills into an oil accumulation (which is under considerable natural *reservoir* pressure), free gas in the *gas cap* above the oil *zone*, expands. This forces the oil to flow into the wellbore, and up the the surface ('*gas-cap* drive'); at the same time, *solution gas* dissolved in the oil comes out of solution, and it too expands, helping to force oil to flow into the wellbore and up to the surface ('*solution gas* drive'). See *water drive.* *

GAS INJECTION - A method of *secondary recovery* in which a gas (usually dry *natural gas* or carbon dioxide) is injected into an oil *reservoir* to increase *reservoir* pressure around the injection well. Some of the gas dissolves in the oil, increasing the oil's ability to flow. The increased pressure and flowing ability of the oil result in increased production from the nearby production wells. *

GAS LIFT - A method of *secondary recovery* similar to *gas injection*, except that the injection well and the production well are both the same well. Dry natural gas is pumped: down through the (*annular*) space between the *casing* and the production *tubing*, and into the *reservoir*. Gas dissolves in oil increasing the oil's ability to flow, and *reservoir* pressure is increased around the well. Pumping stops. Then, as pressure is bled off at the surface, the oil-and-gas mixture (which is lighter than the oil by itself) flows from the *reservoir*, into the production *tubing* in the well, and up to the surface. The procedure of injecting gas and then flowing the oil-and-gas mixture is carried on intermittently ('huff and puff'). The gas is separated from the mixture at the surface, where it is stored for re-injection.

GAS-OIL RATIO (GOR) - The volume of gas produced along with the oil from an oil well, usually described in *MCF* (thousands of cubic feet of gas) per *barrel* of oil. In the most general sense, an oil and gas accumulation might be broadly classified according to its GOR:

GOR (MCF/BBL)	Probable Hydrocarbon
0 - 2,000	Oil, no free gas
1,000 - 20,000	Oil / Gas / Condensate
> 20,000	Dry gas

GENERAL PARTNER - In a *limited partnership*, the general partner is responsible for managing the partnership's activities (and is commonly the party that put the deal together). His liability to the partnership's creditors is unlimited. *

GEOPHONES - Microphones placed on the ground to detect sound waves generated during *seismic* surveying. *

GEOPHYSICIST - A geophysicist applies the principles of **phys**ics to the understanding of **geo**logy. Emphasis is generally on the earth's gravitational and magnetic fields, and the travel of sound waves through the earth. Many U.S. universities offer BS, MS, and PhD degrees in geophysics; the Society of Exploration Geophysicists has over 19,000 members.

GROSS ACRES - The number of *acres* in which one owns a *working interest*. A net *acre*, is a gross acre multiplied by one's *working interest* in it.

GROSS INCOME - Total income from an activity, before *deduction* of (1) items that may be treated as *expenses* (such as *intangible drilling costs*), and (2) allowed tax items (such as *depletion* allowance, *depre-*

ciation allowance, etc). See *taxes due.*

GROSS WELLS - The number of wells in which one owns a *working interest.* A net well, is a gross well multiplied by one's *working interest* in it.

GUARANTEED PAYMENTS - Payments by a partnership to one or more of its partners for services rendered. They are are paid without regard to the partnership's income (as if to a third party). *

HELD BY PRODUCTION (HBP) - Refers to an oil and gas property under *lease*, in which the *lease* continues to be in force, because of production from the property. See also *secondary term.*

HORIZONTAL DRILLING - The new and developing technology that makes it possible to drill a well from the surface, vertically down to a certain level, and then to turn a right angle, and continue drilling horizontally within a specified *reservoir*, or an interval of a *reservoir*.

Four gross systems are characterized as 1) short radius, 2) medium radius, 3) long radius,and 4) ultra-short radius. (For comparison purposes: short radius might be 300 feet or less; long radius might be more than 3,000 feet.)

HOT OILING - Some *crude oils* contain significant paraffin (waxy) *hydrocarbons* which 1) reduce the ability of the oil to flow, and 2) remain liquid, only at relatively high temperatures. Production of this type of oil usually tends to decline rather rapidly (as the paraffins in the oil clog the porosity surrounding the well bore).

Hot oiling is a method of (temporarily) alleviating this situation by using heating equipment and special procedures to increase the temperature in the *reservoir* close to the *borehole*, thereby liquefying the paraffin, and un-clogging the pore spaces.

Alternatives to hot-oiling include experimental production techniques, injecting chemical additives into the reservoir, etc.

HYDROCARBONS - A large class of organic compounds composed of hydrogen and carbon. *Crude oil*, *natural gas*, and natural gas *condensate* are all mixtures of various hydrocarbons, among which *methane* (CH_4) is the simplest. *

IDC - See *intangible drilling costs*.

Horizontal Drilling Technology

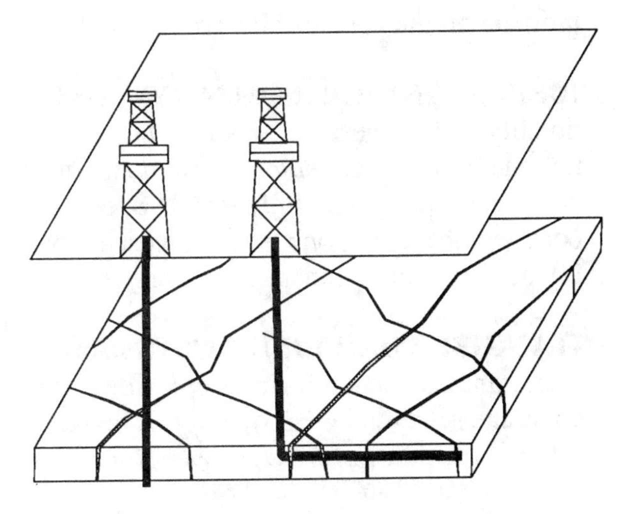

Conventional Well   Horizontal Well

A single horizontal well in a *fractured reservoir* can replace several conventional wells.

Fractures allow *more* oil and gas to flow from the reservoir into the well, *faster!*

INCREASED DENSITY - See *spacing unit.*

INITIAL POTENTIAL - Flow rate measured during the initial *completion* of a well in a specific *reservoir* ('initial daily rate of production'). It may or may not reflect the ultimate producing capability of the well.

INDEPENDENT (OIL COMPANY) - Traditionally, any domestic oil company that is not one of the seven *major* international oil companies. More generally, any U.S. oil company that is not one of the largest 18 or 20 integrated oil companies. See *major*.

INTANGIBLE DRILLING COSTS - An important tax accounting concept. Adequate coverage is beyond the scope of this book.

For complete discussion and examples please refer to MONEY IN THE GROUND.

INVESTMENT TAX CREDIT (ITC) - A credit against income taxes, usually computed as a percent of the cost of investment in certain types of *assets* (specified in tax regulations). It directly offsets *tax liability*. See *taxes due*.

JOINT - A single section of *drill pipe*, *casing*, or *tubing*, usually about 30 feet long.

JOINT OPERATING AGREEMENT (JOA) - A detailed written agreement between the *working interest* owners of a property which specifies the terms according to which that property will be developed.

JUG-HUSTLER - In the field operations of a seismic survey, the *geophones* (jugs) must be placed on the ground in a particular pattern and connected via electrical cables, to a recording computer. After the sound waves are recorded at one location, the jugs have to be picked up, moved to the next location, and again laid out in the specified pattern. The field-hand who does this is a jug-hustler.

KELLY BUSHING (KB) - Part of the *drilling rig*, the Kelly is a long hollow steel bar that connects to the upper end of the *drill string*. It is square or hexagonal in cross-section. The Kelly bushing is a special 'sleeve' in the rotary table through which the Kelly can freely move up and down during drilling.

The depth to a particular *zone* in a well is generally measured from the KB (Kelly bushing), which may be anywhere from about 5 to 50 feet above ground level (depending on the type and size of *drilling rig* being used). *

LAG TIME - The time it takes for *cuttings* to be carried ('*circulated*') from the bottom of the *borehole* up to the surface by the *mud* system. It increases with the depth of the *borehole*, ranging up to several hours. See *circulation*, *bottoms up*.

LANDMAN - An oil company employee or agent who negotiates the purchase of *leases*, cures defects in the *title* to property leased, and assists the oil company in complying with government regulations and reporting procedures. (Some U.S. universities offer a BA degree in Petroleum Land Management, PLM; there are over 12,000 members of the American Association of Petroleum Landmen.)

LAW OF CAPTURE - A legal concept on which oil and gas law in some states is based: since *petroleum* is liquid, and hence mobile, it is not owned until it is produced ('captured').

Oil and gas law in other states is based on the concept that *petroleum* is a mineral and is treated the same as solid minerals (such as coal).

Consider this analogy: During duck-hunting season, FARMER FRED shoots a duck as it flies over his wheat field. His shot hits the duck, wounding it, but the duck manages to continue to flap along, on its course. It eventually falls from the sky, dead, in LANDOWNER LARRY's front yard, several miles away. Who owns the duck?

(Oil and Gas) **LEASE** - A contract by which the owner of the *mineral rights* to a property conveys to another party, the exclusive right to explore for and develop minerals on the property, during a specified period of time. The conveying party is the 'lessor'; the recipient is the 'lessee'.

The terms of a lease are typically negotiated between a *landman* representing an oil exploration company, and the owner of the *mineral rights* of a property: the *bonus*, *royalty*, and *delay rentals* to be paid to the lessor, and the *primary term*, and the *secondary term* of the lease, among various other terms. *

Please refer to MONEY IN THE GROUND for a complete discussion of oil and gas leases.

LEASE or **SUBLEASE** (classification as) - Any transaction in which the owner of operating rights in a property (for example, a *working interest*) assigns all or a portion of these rights to any other party, either for (1) no immediate consideration or (2) for cash (or equivalent), and retains a continuing non-operating interest in production (such as an *override*). If AL owns a 100% *working interest* in a property, and assigns all of it to BOB, retaining a 5% *override*, the transaction would be a sublease. (If AL had retained a 10% *carried working* interest, instead of an *override*, the transaction would not be a sublease.) See *sale*. *

LEASE HOUND - Someone who goes out and aggressively acquires oil and gas leases from the landowner, and then turns around and sells or trades them to an oil company planning to drill a well in the area.

LIFTING COSTS - The expenditures involved in lifting (pumping) oil from a producing *reservoir* in a well, up to the surface. Included in *lease* operating (production) expenses.

LIMESTONE - *Sedimentary* rock largely consisting of *calcite* (calcium carbonate). On a world-wide scale, limestone *reservoirs* probably contain more oil and gas reserves than than all other types of *reservoir* rocks combined.

LIMITED PARTNER - In a *limited partnership*, a partner whose liability is limited to the amount of his investment in the partnership (plus any *assessments* and his share of

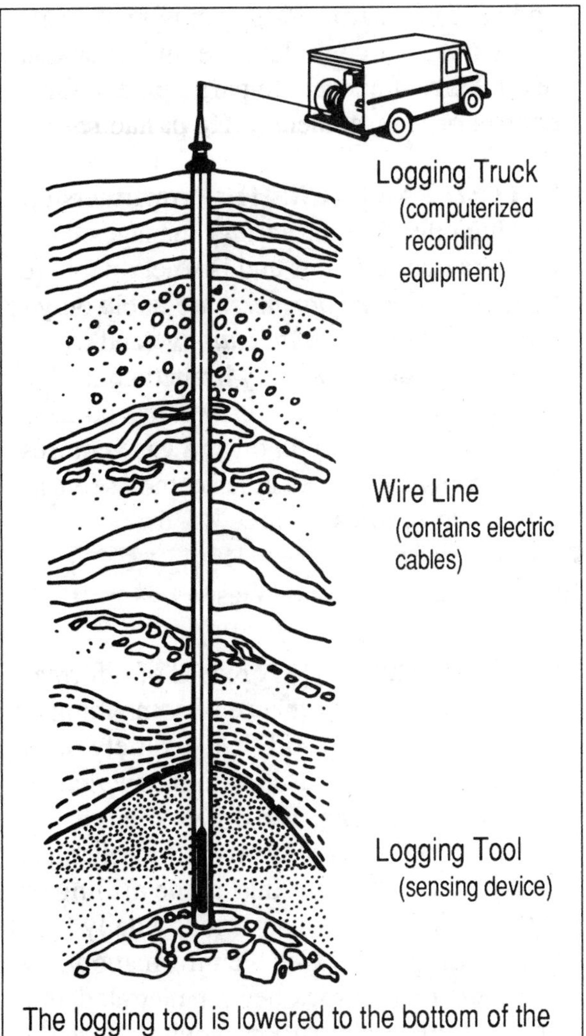

The logging tool is lowered to the bottom of the well. Measurements are recorded as the tool is pulled back up to the surface.

undistributed partnership earnings). A limited partner could lose the limitations on his liability if he were to participate in the control or management of the partnership. *

LIMITED PARTNERSHIP - A partnership in which the *general partner* manages the partnerships activities and is solely liable for them. The *limited partners* are liable only to the extent of their *contributions* (and *assessments*), and have only limited control over policy decisions. It is the most common vehicle for investing in oil and gas ventures used by investors who are not, themselves, oil industry professionals. *

LNG - Liquefied Natural Gas.

LOG, LOGGING - (1) A record of information about the performance of some thing or some process. In the oil field a variety of logs are used in the drilling and *completion* of oil and gas wells. The driller keeps a 'driller's log', which records all mechanical operations undertaken during the drilling of the well. The *mud logger* keeps a daily 'mud log' recording all available information about the type of rock being penetrated and any possible hydrocarbon *shows*.
(2) Wireline logging is the process of lowering a sensing device down into the *borehole* on

the end of an electric cable (a line made of wire). Wireline logs record measurements of various parameters providing information about the type of rock penetrated, the fluids contained in the rock, or the condition of the borehole. Parameters that can be measured include: diameter of the borehole (*caliper log*), electrical resistivity of the rock penetrated (resistivity or electric log, E-log), density of the rock penetrated (density log and neutron log), acoustic properties of the rock penetrated (sonic or acoustic log), natural gamma-ray radiation of the rocks penetrated (gamma ray log, GR), inclination or dip of the rocks penetrated (*dipmeter log*), borehole temperature (temperature log).

Until fluids have been recovered, wireline logs may be the primary source of information on the quantity and quality of hydrocarbons in the rocks penetrated by the borehole. After drilling the well, the decision to spend additional money to attempt to *complete* the well to produce oil or gas is often based exclusively on the analysis of wireline log data. *

LOST CIRCULATION - See *circulation*.

MAJOR (OIL COMPANY) - Traditionally used to refer to the integrated international companies ('the seven sisters'): British Petroleum, Exxon, Gulf, Mobil, Shell, Standard of California (Chevron), and Texaco. (Gulf has since been acquired by Chevron.) Today 'major' may be used more loosely to include any of the twenty or so largest integrated oil companies. See *Independent*.

MANAGEMENT FEE - In oil and gas *limited partnerships*: a fee paid by the *limited partners* to the *general partner* for services he provides in the management of the partnership (according to the partnership agreement). The amount of such fee is spelled out in the partnership agreement. It may be specified as a *guaranteed payment* equal to a percentage of *capital contributions*, or specified as a fee per well, a fee per *leased* property, etc.

MAXIMUM EFFICIENT RATE (MER) - The experimentally determined flow-rate that permits optimum recovery of oil or gas from a well. Before a well is put on commercial production, different flow rates are tested for their effect on the composition of the oil-gas-water mixture produced, and on the *depletion* of natural *reservoir* pressure.

MCF - Thousands of cubic feet measured at standard temperature (60° F) and pressure (14.65 psi). In the U.S., MCF is the most common unit of measure for volumes of *natural gas*. ('M' is the Roman numeral for 1,000; MMCF indicates millions of cubic feet). See *BCF*. *

MEASUREMENT WHILE DRILLING (MWD) - The still-developing technology of measuring various *downhole* parameters during the drilling of a well (without having to pull the *drill string* out of the hole in order to run *wireline logs* or a directional survey, as in conventional drilling procedures).

Data recorded by downhole sensors is transmitted up to the surface during the drilling process, by various electric or physical systems.

These MWD systems play a vital role in the new and developing technology of *horizontal drilling*.

METHANE - The simplest of the *hydrocarbons*, CH_4. It is a colorless, odorless gas which generates about 1012 *Btu* of heat energy per 1,000 cubic feet of gas (*MCF*), when burned.

MIGRATION - The movement of oil and gas through layers of rock deep in the earth. *

MINERAL ACRE - A full *mineral interest* in one *acre* of land.

MINERAL INTEREST ('MINERALS', 'MINERAL RIGHTS') - The ownership of all rights to gas, oil, or other minerals as they naturally occur in place, at or below the surface of a tract of land. Ownership of the minerals carries with it the right to make such reasonable use of the surface as may be necessary to explore for and produce the minerals. Only the mineral owner (or *fee owner*) may execute an oil or gas *lease* conveying his interest in a tract of land. See *severance*. *

Please refer to MONEY IN THE GROUND for complete discussion of mineral interests.

MUD (DRILLING MUD) - A fluid mixture of clays, chemicals, and weighting materials suspended in fresh-water, salt-water, or diesel oil. See *circulation.*

It (1) cools and lubricates the *drill bit*, (2) carries *cuttings* to the surface, (3) maintains the required pressure at the bottom of the hole, and (4) coats the inside of the *borehole* with a sort of plaster called 'mud cake' which helps prevent the walls from caving into the *open hole*.

'Mud weight' is the density of the mud measured in pounds per gallon (ppg). It can be controlled at the surface by the addition of various substances, such as barite (barium sulfate - $BaSO_4$), a heavy mineral 4.5 times as dense as water:

8.3 ppg = the density of water
15 ppg = 2 x the density of water
20 ppg = 2.4 x the density of water

The pressure exerted by the column of mud in the borehole must be greater than natural pressures in rocks likely to be encountered during drilling. Drilling with a mud weight that is too low could result in a *blowout.*

MUD ENGINEER - A technician responsible for proper maintenance of the *mud* system.

MUD LOGGER - A technician who uses chemical analysis, microscopic examination of the *cuttings*, and an assortment of electronic instruments to monitor the *mud* system for possible indications of *hydrocarbons* ('*shows*'). See *log*.

MULLET - Don't you be one! (An unflattering reference to someone willing to entertain putting money in an oil and gas venture without actually understanding the specifics of the deal; a sucker.)

MULTIPLE COMPLETION - See *dual completion.*

MWD - See *measurement while drilling.*

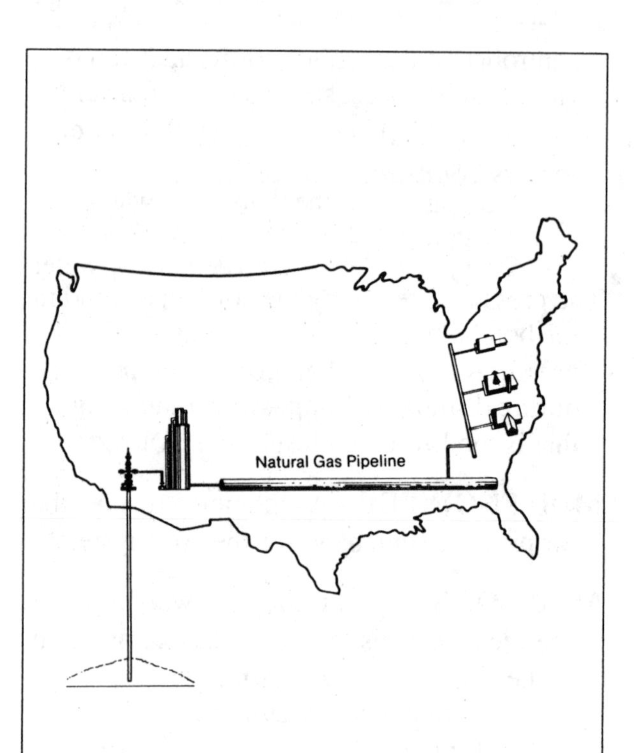

There are more than 250,000 miles
of natural gas pipelines in the U.S.

GAS: from Reservoir to Consumer

NATURAL GAS - A mixture of gaseous *hydrocarbons* formed naturally, in the earth. Most natural gases contain *methane* as the primary component, mixed with ethane C_2H_6, propane C_3H_8, butane C_4H_{10}, pentane C_5H_{12}, and/or hexane C_6H_{14}. Non-hydrocarbon gases such as carbon dioxide, helium, nitrogen, and hydrogen sulfide may also be present.

NET ACRES - See *gross acres.*

NET PAY (NET FEET OF PAY) - The aggregate thickness (in feet) of the *pay zone* likely to contribute to production from a well. The footage count excludes portions of the *reservoir* having *porosity* lower than a specified cut-off value (5%, for example) and having water *saturation* greater than a specified cut-off value (45%, for example). Cut-off values are established on the basis of historical production data in the same or a similar *reservoir*, in the same or a similar geologic setting.

NET PROFITS INTEREST - A share of gross production from a property that is *carved out* of a *working interest*, and is figured as a function of net profits from operation of the property.

Depending on the specifics of the arrangement, a net profits interest might be required to bear certain expenses of development and operations, but would neither be obligated to advance expenses nor liable for losses (as in the case of the *working interest*).

NRI - See *net revenue interest.*

NET REVENUE INTEREST (NRI) - The percentage of revenues due an interest holder in a property, net of royalties or other burdens on the property. Assume LANDOWNER *leases* his *mineral rights* to OILMAN. LANDOWNER retains a *royalty* of 1/8 (= 12.5%); his net revenue interest is 12.5%. OILMAN'S net revenue interest would be 87.5% (= 100% - 12.5%). *

Please refer to MONEY IN THE GROUND for more detailed discussion and examples of net revenue interest.

NET WELLS - See *gross wells.*

OFFERING MEMORANDUM - A legal document provided to potential investors in a venture (such as an oil and gas *limited partnership*), describing the terms under which the investment is being offered.

OIL COLUMN - The vertical height (thickness) of an oil accumulation above the oil-water contact. In commercial oil fields, oil columns can range from 5 feet to (rarely!) several hundred feet. See *gas column.* *

OIL GRAVITY - The density of liquid *hydrocarbons*, generally measured in degrees *API* (American Petroleum Institute). Degrees *API* are inversely related to specific gravity: Degrees *API* = 141.5/Sp Gr @ 60° F minus 131.5

The lighter the oil the higher the *API* gravity: A tar with an *API* gravity of 8° *API* would be heavier than water; motor lubricating oil is typically about 26° *API*; gasoline is about 55° API.

OPEN HOLE - Refers to a *borehole*, or the portion of a *borehole*, in which *casing* has not been set. See *cased* hole.

OPERATOR - The *working-interest* owner responsible for the drilling, *completion* and production operations of a well, and the physical maintenance of the *leased* property. Responsibilities of the operator and other *working-interest* owners are enumerated in the *joint operating agreement.*

ORGANIZATION COSTS - Direct costs incurred in the creation of a new business organization such as an oil and gas *limited partnership*: legal and accounting fees, organizational meetings, printing costs. In the calculation of partnership income, these are *amortizable capital* expenditures, generally *amortized* by the partnership over a period of 60 months from the start-up of the business.

OVERRIDING ROYALTY ('OVERRIDE', 'ORRI') - A revenue interest in oil and gas, created out of a *working interest.* Like the lessor's royalty, it entitles the owner to a share of the proceeds from gross production, free of any operating or production costs (but net of state production and *severance* taxes). It terminates when the lease expires. See *royalty*. *

Please refer to MONEY IN THE GROUND for complete discussion and examples of overriding royalties.

PACKER - A flexible rubber sleeve that is part of a special joint of pipe (tool). During certain (non-drilling) activities, the tool may be added to (inserted in) the *drill string*. When the *drill string* is lowered into the *borehole*, the packer can be expanded (from operations at the surface), to temporarily block off a portion of the *annulus* of the borehole (as may be done during *drill stem tests* or *squeezing* operations.

PAD - Refers to the "Petroleum Administration for Defense", which groups the 50 states into five regional districts for administrative and reporting purposes.

PAY ZONE or **PAY** - An interval of rock, from which oil or gas is expected to be produced in commercial quantities. *

PAYOUT - In the case of an oil and gas investment, the point at which costs of exploration and development (or amount of the investment) are recouped out of net revenues received from production.

In many oil and gas *limited partnerships* (and standard deals within the oil industry) there may be a re-allocation of revenues or cash flow among the participants in the investment, after payout has been reached. *

PERFORATING GUN - An instrument lowered at the end of a *wireline* into a *cased well.* It contains explosive charges that can be electronically detonated from the surface. When it is at the level of the *pay zone*, the bullets are shot into the *reservoir* rock, piercing the *casing* (and surrounding cement sheath). The resulting holes allow rock fluids to flow from the *pay zone* into the *borehole.* *

PERMEABILITY - The measurable capacity of a rock to allow a fluid to flow through it. Along with *porosity*, it is one of the two most critical properties of the reservoir rock. To be commercially producible, oil or gas must be able to flow from the *reservoir* rock into the well. *

PETROLEUM - Crude oil. Naturally occurring liquid *hydrocarbons* from which gasoline, kerosene, and countless other 'petrochemicals' are produced. Non-hydrocarbon compounds (often containing sulfur and nitrogen) may also be present. In the broadest sense, petroleum refers to all naturally occurring *hydrocarbons*, including *oil*, *natural gas*, *condensate*, and their derived products.

PETROLEUM ENGINEER - A term including three areas of specialization: DRILLING engineers specialize in the drilling, workover, and *completion* operations; PRODUCTION engineers specialize in studying a well's characteristics and using various chemical and mechanical procedures to maximize the recovery from the well; RESERVOIR engineers design and execute the planned development of a *reservoir*. Many U.S. universities offer BS, MS, and PhD degrees in petroleum engineering; the

Society of Petroleum Engineers has over 50,000 members.

PETROLEUM GEOLOGIST - A geologist who specializes in the exploration for, and production of, petroleum. Many U.S. universities have departments which offer BS, MS, and PhD degrees in petroleum geology; the American Association of Petroleum Geologists has about 38,000 members.

PLUG - An object (usually of cast iron or cement) set in a *borehole* to block the passage of fluids. During various operations in the *completion* or the *abandonment* of a well, it may be necessary to isolate certain sections of the well. Some plugs can be drilled through ('drilled out') after the operation is completed.

PLUG BACK - To block off the lower section of the *borehole* by setting a *plug*, in order to perform operations in the upper part of the hole.

PLUGGED & ABANDONED (P&A) - This expression refers to setting cement plugs in an unsuccessful well (a 'dry hole') or a depleted well, before *abandoning* the well.

POOLING - When tracts of land are smaller

than the *spacing unit*, the separately owned *working interests* (of these tracts) may be brought together ('pooled') in order to obtain a drilling permit.

In states where pooling of separately held *working interests* is required ('compulsory pooling' or 'forced pooling' – notably Oklahoma and Arkansas): in order to obtain a drilling permit, any *working- interest* owner can initiate a forced pooling order.

A force-pooled *working-interest* owner typically has 3 options: (1) participate in the well by paying his proportionate share of the costs, (2) *farm-out* his interest to the pooling party, or (3) sell or *sublease* his interest in the proposed well to a third party.

POOR-BOY - To 'poor-boy' a well means to try to make do without adequate financing: to borrow money, sell interests, make trades, borrow equipment, or otherwise somehow get ahold of enough money or credit, to get the well drilled.

POROSITY - The percentage (by volume) of holes or voids in a rock. Commercially productive *reservoir* rocks typically have porosities ranging from about 5% - 35%. The higher the porosity, the more oil or gas that can be contained in the pore spaces, the better the quality of the *reservoir* rock. *

PRESENT NET VALUE - The present value of the dollars (income, or stream of income) to be received at some specified time in the future, discounted back to the present at a specified interest rate.

PRIMARY RECOVERY - Production in which oil moves from the *reservoir*, into the *wellbore,* under naturally occurring *reservoir* pressure. In a flowing well, the *reservoir* pressure is great enough to force the oil up to the surface. In a pumping well, oil is *artificially lifted* to the surface by pumping equipment. *

PRIMARY TERM - The basic period of time during which a *lease* is in effect. It may range from 6 months to more than 10 years, but is perhaps most commonly 3 - 5 years. One of the negotiable provisions of a *lease.* See *secondary term.*

PRIVATE PLACEMENT OFFERING - A *securities* (investment) offering not intended for the general public. By meeting certain criteria, such an offering may qualify for exemptions from registration with (1) the Securities and Exchange Commission of the Federal government and/or (2) the *securities*-regulating agencies of the various State governments involved. See *public offering.*

PRODUCTION (OIL) PAYMENT - A specified share of gross production from a property that may be *carved out* of a *working interest*, or *retained* by the transferor of a *working interest*. It is limited: (1) to a specified period of time, (2) until a specified amount of money is received, or (3) until a specified quantity of minerals has been received.

PROPPANT, PROPPING AGENT - See *fracturing*.

PROSPECT - The hypothesis that a naturally occurring, commercially exploitable accumulation of oil or gas exists, at a clearly defined underground location. It is described by one or more geologic maps. A single well should be sufficient to test the hypothesis. Surface area of a prospect could range from a ten-acre tract large enough for only a single well, to a tract covering many square miles and requiring dozens of wells to fully develop its reserves.

Prospects are typically named for some nearby geographic or geologic feature, but the name may simply reflect the whim of the moment. A well is commonly named for the lease on which it is drilled. ("The J. Jones No. 1 was drilled on the Bear Creek prospect, last year.") *

Please refer to MONEY IN THE GROUND for a complete discussion of oil and gas prospects.

PUBLIC OFFERING - A *securities* (investment) offering intended for sale to the general public. It must be registered with (1) the the Securities and Exchange Commission of the Federal government and (2) the *securities*-regulating agencies of the various States in which it will be offered. See *private placement.*

QUITCLAIM DEED - A document by which one party (grantor) conveys *title* to a property, by giving up any claim which he may have to *title* (although he does not profess that his claim is necessarily valid). See *deed.*

RECAPTURE - Repayment to the Federal government of tax benefits (such as income tax *deductions* or *tax credits*) that may become necessary upon a development that is contrary to the assumptions on which the the tax benefits were originally based.

Assume that an investor in an oil and gas property enjoyed the benefit of *intangible drilling cost deductions* against his *gross income.* If he later sells the property at a gain, some of those *deductions* may be subject to recapture.

RE-ENTRY - Assume that a well was *abandoned*, but subsequent drilling and production in the area suggests that a potential pay zone in the well was missed or 'passed over'. Instead of drilling a second well to evaluate the *zone of interest*, a company might drill out the cement abandonment *plugs* in the original *borehole* to test the potential zone. Such a 're-entry' can be a rather risky operation, but if successful, the cost may be much less than drilling a new off-setting well.

RESERVES - The amount of oil or gas in a *reservoir* currently available for production, usually described as *barrels* of oil, or *MCF* (thousands of cubic feet) of gas, attributable to a well, to a property, or to an entire field. The term should always be qualified by an adjective, since there are many ways of estimating the reserves.

'**Reserves in place**' describes the amount of oil or gas physically contained in the *reservoir*. The amount of reserves that can actually be gotten out of the ground (can be recovered) may be only 25% to 30% of the reserves in place.

'**Recoverable reserves**' is an estimate of the amount of oil or gas that can be produced from the *reservoir* in the future. (No one can know what the actual recoverable reserves are until after they have been produced.)

Proved Reserves: the amount of reserves that are considered to be recoverable under existing technology and economic conditions (based on available geologic and engineering data).

Proved developed producing reserves: proved reserves recoverable from currently completed intervals in existing wells using existing facilities. These are the numbers a banker might want to see in considering an application for a loan.

Proved behind pipe reserves: reserves in zones behind the casing in an existing well (which might be produced if the well were re-completed in those zones).

Proved undeveloped reserves: reserves which could be produced, but drilling new wells, deepening existing wells, or secondary recovery methods would be required.

Estimates of recoverable reserves classified as 'probable future reserves' and 'possible future reserves' may include a wide variety of geologic and engineering assumptions. Estimates of 'undiscovered recoverable reserves' refer to reserves outside of known accumulations and are, at best, a guess developed from geologic and statistical game theory. See *EUR*.

RESERVOIR - Any rock having enough *porosity* and *permeability* to contain appreciable *hydrocarbons*. Most commercial reservoirs are in *sandstone* and *limestone* (limestone/dolomite), rarely in shale or igneous rock.

(Electrical) **RESISTIVITY** - The ability of a substance to impede the flow of electricity through it. Variations in the resistivity of different rocks depends largely on the fluids contained in the pores of the rocks: pure oil, pure gas, and fresh water each have very high resistivity, while salt water can have very low resistivity. Most dry rocks do not

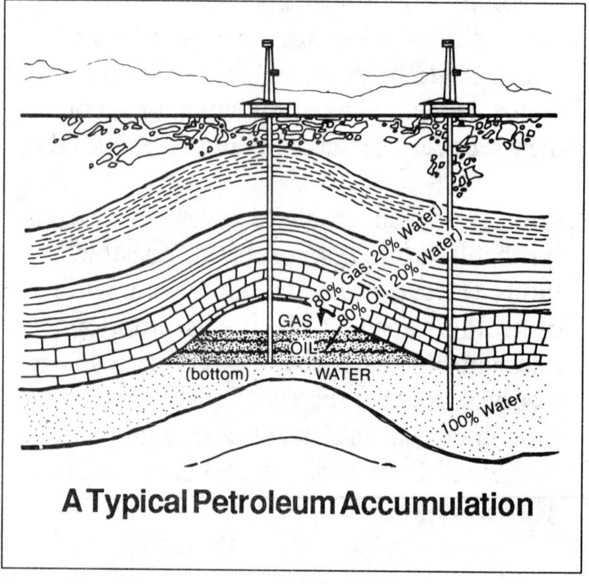

A Typical Petroleum Accumulation

conduct electricity (and thus have infinitely high resistivity).

RETAINED INTEREST - A fractional interest reserved by the owner of a whole interest when the balance of the whole interest is transferred (conveyed) to another party. See *carved out* interest.

REVERSIONARY INTEREST - An interest in a well or property that becomes effective at a specified time in the future or on the occurrence of a specified future event. See *back-in*.

Please refer to MONEY IN THE GROUND for more detailed discussion and examples of various reversionary interest deals.

RISK - Literally: 'the possibility of loss or injury'. In oil exploration, a level of uncertainty is associated with the various possible outcomes of the undertaking. Risk usually refers to a numerical estimate of the likelihood of the occurrence of these various possible outcomes.

Drilling a well in an area where: (1) many wells have already been drilled, and (2) the results of previous drilling have been thoroughly analyzed and evaluated, can be a relatively low-risk proposition:

- The technology for drilling wells in this

area has already been established.

- There are numerical data from which it is possible to estimate the statistical probability of finding oil or gas.
- Production statistics allow estimates of reserves per successful well.
- Most importantly, there are abundant geologic data from which to synthesize a scientific geologic understanding of the sub-surface (which is the key to beating the purely numerical odds of finding oil and gas by random selection of drilling locations).

By contrast, drilling a well at the South Pole would be a very high-risk proposition, indeed.

ROUGHNECK - One of the two or three field-hands on the derrick floor during drilling operations, whose job revolves around breaking out the *drill pipe*, making-up connections, and stacking *drill pipe*. It is a tough, physically demanding, and often dangerous job (not for lightweights or sissies). See *weevil*.

ROUSTABOUT - A semi-skilled hand who looks after producing wells and production facilities. The name reflects the great diversity of general maintenance jobs that must be done. The position is frequently offered to former *roughnecks*, who due to age or injury, may no longer be fit for the strenuous work on a *drilling rig* during drilling operations.

ROYALTY - A right to oil and gas (or other minerals) as they naturally occur in place. It entitles the owner to a specified share of oil and gas produced from a property, but bears no costs or responsibility (or rights) for the development of the property. Although he does not pay operating and production expenses the royalty owner does pay state *severance tax* (is subject to Federal income tax); depending on the provisions of the *lease*, he may also pay handling costs or costs required to get the product from the wellhead to a pipeline connection or refinery.

If the *minerals owner leases* his land and retains a 1/8 royalty, he is entitled to receive 1/8 of the gross production from that property, at no cost to himself (either in kind or in value, as agreed upon in negotiation of the *lease*).

In some foreign countries a royalty interest may consist, not of the right to a specified fractional share of oil and gas produced from a property (as in the U.S.), but merely of a claim against the *working interest* owner for oil and gas produced. In such foreign jurisdictions, the royalty owner's claim may only be one among numerous contractual claims against the working interest owner. *

Please refer to MONEY IN THE GROUND for complete discussion of royalties.

ROYALTY ACRE - The (*mineral owner's*) *royalty* on one *acre* of *leased* land.

SALE (classification as) - When cash (or equivalent) is received as consideration, three types of transactions might be classified as a sale:

- the owner of any kind of oil and gas interest assigns all of his interest to another party (or assigns a fractional part and retains a fractional part of that same identical interest),
- the owner of a *working interest* assigns a continuing non-*operating* interest (such as an *override*), but retains the *working interest*,
- the owner of any type of continuing oil and gas interest, assigns hat interest, and retains a non-continuing interest in production (a *production payment*).

Classification of the transfer of oil and gas interests as a sale, or as a *lease* or sublease, can have significant economic consequences both to the transferor (affecting the rate at which his income is taxed — ordinary rates or capital gains rates), and to the transferee (affecting his entitlement to *depletion allowance* and the methods by which it may be calculated — percentage *depletion* or cost *depletion*). *

SANDSTONE - Rock composed mainly of sand-sized particles or fragments of the mineral quartz. Individual grains often occur naturally "glued" together by the mineral *calcite*. Because quartz grains are rigid, the fabric of the rock will withstand tremendous pressures, without being compacted. In this sense, sandstone is very different from *shale*.

SATURATION - (1) Water-Saturation: the fluid contained in the pores of an oil and gas reservoir rock usually consists of a mixture of water plus oil and/or gas. Water-saturation is the percent of water contained in this mixture. A low water-saturation (20% for example) implies a high concentration of hydrocarbons (approximately 100 - 20 = 80%) and suggests that the rock will produce oil or gas. A high water-saturation (greater than 85%) implies that the rock is likely to produce only water.
(2) In a *reservoir* containing oil, the crude oil is said to be saturated when it contains as

much dissolved *natural gas* as is physically possible. Any excess gas would accumulate above the oil as a *gas cap*.

SECONDARY RECOVERY - After *primary recovery* operations have taken their course, various operations may be undertaken to increase the amount of oil that can be by normal methods of flowing and pumping.

The second stage consists of efforts to increase production by addressing the condition of the *reservoir*. Typical operations involve forcing gas ('*gas injection*'), or water ('water flooding') into the *reservoir*. This serves to re-pressurize the *reservoir*, which allows recovery of more oil than would be possible from *primary recovery*.

Tertiary (third stage) recovery includes further efforts to recover additional oil from the *reservoir*, by altering the physical characteristics of the oil itself - reducing viscosity, or reducing surface tension. These may be largely experimental procedures, like the injection of CO_2, detergent-like fluids, steam, or chemically treated water into the reservoir; or injecting air (oxygen) into the reservoir and burning some of the oil in place, to raise the temperature of the oil and improve its ability to flow. *

SECONDARY TERM - The period of time

that a *lease* is automatically extended beyond the *primary term,* as long as there is active drilling or production. See *primary term, lease.*

SECTION - A square tract of land having an area of one square mile (= 640 *acres*). There are 36 sections in a *township.* *

SECURITIES - Securities are commonly thought of as stocks and bonds. As defined by the Securities Act of 1933, however, securities include any certificate of interest or participation in any profit sharing agreement, investment contract, or fractional undivided interest in oil, gas, or other mineral rights.

SECURITIES ACT OF 1933 - Establishes requirements for the disclosure of information for any interstate offering and sale of *securities.*

SECURITIES EXCHANGE ACT OF 1934 Established the Securities and Exchange Commission which regulates the activities of *securities* markets.

SEDIMENTARY ROCK - Rock that is naturally formed from fragments of other rocks. These fragments result from mechanical abrasion ('weathering') of pre-existing rock, and are transported by water, ice, and air. Sedimentary rocks that are important in terms of petroleum include *sandstones* and *limestones*, which are often *reservoir* rocks, and *shale* which may be a *source rock.* *

SEISMIC SURVEYING - The procedure of sending pulses of sound from the surface, down into the earth, and recording the echoes reflected back to he surface. By making assumptions about the speed at which sound travels through the various layers of rock, it is possible to estimate the depth to the reflecting surface. It then becomes possible to infer the structure of rocks deep below the earth's surface. *

SELLING EXPENSES - Expenses incurred in marketing interests in *securities* (and commonly paid out of the investor's *capital* investment).

SEVERANCE - The owner of all rights to a tract of land (the *fee simple* owner) can sever the rights to his land (vertically or horizontally). In horizontal severance, for example, if he chooses to sell all or part of the *mineral rights*, two distinct estates are created: the *surface rights* to the tract of land and the *mineral rights* to the same tract. The two

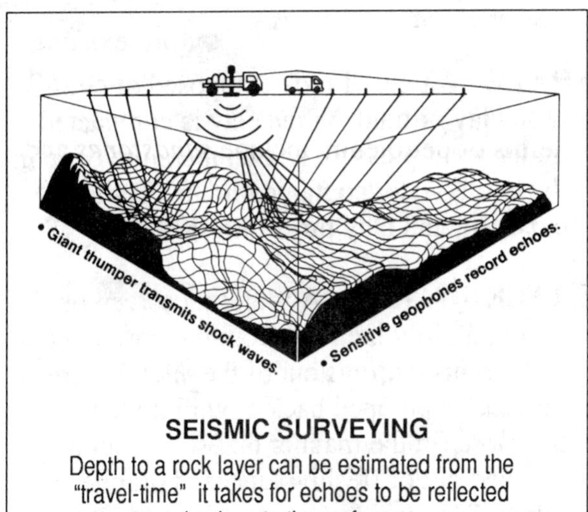

SEISMIC SURVEYING

Depth to a rock layer can be estimated from the "travel-time" it takes for echoes to be reflected back up to the surface.

estates may change hands independently of each other. Severed *mineral rights* may be restricted as to mineral type, or limited by depth, (in which case the landowner retains the rights to minerals other than those severed, and to depth intervals other than those severed). *

SEVERANCE TAX - A tax paid to the state government by producers of oil or gas in the state. It may be specified either as a percent of the oil or gas taken ('severed') from the earth, or as a dollar amount per *barrel* of oil or per thousand feet of gas (*MCF*) produced (also called 'production tax'). *

Please refer to MONEY IN THE GROUND for

severance tax rates, by state.

SHALE - A type of rock composed of common clay or mud. When clay is compacted under great pressure and temperature deep in the earth, water contained in the clay is expelled, and clay turns into shale.

SHALE SHAKER - A vibrating screen or sieve (slightly inclined from the horizontal) that strains *cuttings* out of the *mud* before the *mud* is pumped back down into the *borehole* (in the manner that seeds can be strained out of freshly-squeezed orange juice). See *circulation.* *

SHARING ARRANGEMENT - (1) Among oilmen: an arrangement whereby a party contributes to the acquisition, or exploration and development, of an oil and gas property, and receives as compensation, a fractional interest in that property.
(2) In a *limited partnership*: the basis or formula for the allocation of costs and revenues between the *limited partners* and the *general partner*.

SHOW - An indication of oil or gas observed and recorded during the drilling of a well. Based on experience, a geologist would rate a show as 'good', 'fair', or 'poor', etc. *

SHUT-IN - To stop a producing oil and gas well from producing.

SHUT-IN ROYALTY - A special type of *royalty* negotiated in the *leasing* of a property. It normally pertains to gas production. If a commercially producible gas well is shut-in due to the lack of a gas market, the *lease* will remain in effect so long as the *working-interest* owner (lessee) pays the specified shut-in *royalty* to the *mineral-rights* owner (lessor).

SIDE TRACK - When *fishing* operations have been unable to recover an object in the hole that prevents drilling ahead, the borehole can often be drilled around the obstacle in the original hole. This 'deviated' hole is called the side track hole. Although it is somewhat risky, drilling a side track hole may be much less costly than starting a new *borehole* from the surface. See *whipstock.*

SLANT HOLE - See *directional drilling*.

SOLUTION GAS - *Natural gas* that is dissolved in the *crude oil* in a *reservoir*. It comes out of solution as pressure is reduced when the oil is produced and flows up to the surface (the way CO_2 bubbles out of a freshly opened can of soda pop). *

SOUR CRUDE, SOUR GAS - Oil or natural gas containing sulfur compounds, notably hydrogen sulfide (H_2S) a poisonous gas. When dissolved in water, H_2S forms a weak solution of sulfuric acid. Over time, this can corrode and destroy metal pipes and equipment. Sweet crude and sweet gas do not contain these sulfur compounds, are less damaging to equipment, and hence generally bring a better price than sour crude and sour gas.

SOURCE ROCK - *Sedimentary* rock, usually *shale* (or *limestone*) containing organic carbon (plant and animal remains) in concentrations as high as 5-10% by weight. After being subjected to high temperatures and pressures during millions of years deep in the earth, the organic material is transformed to liquid or gaseous *hydrocarbons* (which make up *petroleum*). Usually these *hydrocarbons* are (naturally) expelled from the source rock, flowing into nearby *porous* rocks. Because of their natural buoyancy with respect to water, they tend to *migrate* upward, to emerge at the surface as an oil or gas seep unless they become trapped along the way. Oil and gas fields are accumulations of such trapped hydrocarbons.

Some oil shales are believed to represent source rocks in which oil was formed, but from which it was never expelled.

SPACING UNIT - The size (amount of surface area) of a parcel of land on which only one producing well is permitted to be drilled to a specific *reservoir*. It is intended that the single well should drain all (or nearly all) of the recoverable oil or gas from that part of the *reservoir* that lies within the spacing unit. State agencies regulate the size of the spacing unit for different *reservoirs*, to facilitate efficient exploitation of oil and gas from them.

Oil's ability to flow, increases with temperature (which increases with depth). Gas flows more readily than oil. The size of the spacing unit for a particular *reservoir* is set according to its depth and production characteristics, and whether the production is oil or gas.

'Increased density' means the spacing unit is reduced: 2 wells (instead of one) per 640 *acres*, amounts to 1 well per 320 *acres*. *

SPOT MARKET - A short-term contract (typically 30 days) for the sale or purchase of a specified quantity of oil or gas at a specified price.

SPUD - To spud a well means to start the initial drilling operations.

SQUEEZE - The procedure of pumping a slurry of *cement* into a particular space in the *borehole* (often the *annulus* between the *borehole* and the *casing*), so that the *cement* will solidify to form a seal.

STOCK TANK BARREL (STB) - A barrel of oil at the earth's surface. Recoverable oil is measured in stock tank barrels. Due to the complex interplay between pressure and temperature, the volume of a mixture of oil and gas at reservoir conditions is usually different from the volume of the same amount of the mixture at the surface. (In the

Before State Regulation of Spacing Units
Healdton Field in Southern Oklahoma (in 1914)

Photo Courtesy of Mac McGalliard and McGalliard Historical Collection, Ardmore Public Library, Ardmore Okla.

early days oil from a new discovery was sometimes collected in the tanks or ponds on a farm that hold water for livestock.)

STRIPPER OIL WELL - An oil well capable of producing no more than 10 *barrels* of oil per day. (At $20/*barrel*, such a well could produce gross revenues of as much as $200 per day). A STRIPPER GAS WELL is a well that produces an average of less than 60,000 cubic feet of gas per day, measured over a 90-day period. (At $2.00/*MCF* such a well could produce gross revenues of as much as $120 per day).

About 73% of the 619,000 producing oil wells in the U.S. in 1987 were stripper wells; combined, they accounted for about 15% of the nation's oil production.

SUBSCRIPTION - The manner by which an investor participates in a *limited partnership* through investment.

SUBLEASE - See *lease* and *sale*.

SUPERVISORY FEE - Analogous to a *management fee* in an oil and gas *limited partnership*, it is paid by the partnership to the *general partner* for direct supervision of mechanical operations at the well site.

SURFACE RIGHTS - Surface ownership of a tract of land from which the *mineral rights* have been severed. Surface rights include the same full use and enjoyment rights that belong to *fee simple* ownership, except that surface possession is subject to the *mineral* owners right of access to the land for the purpose of extracting his minerals. *

Please refer to MONEY IN THE GROUND for discussion of surface and mineral rights.

SWAB - A hollow rubber cylinder with a flap (check valve) on the bottom surface. It is lowered below the fluid level in the well. This opens the check valve allowing fluid into the cylinder. The check valve flap closes as the swab is pulled back up, lifting oil to the surface.

SWEET CRUDE/GAS - See *sour crude*.

SYNDICATION EXPENSES - Expenditures incurred by a partnership in connection with issuing and marketing its interests to investors: legal fees of the issuer for *securities* and tax advice, accounting fees for audits and other representations included in the offering memorandum, registration (with the Securities and Exchange Commission of the Federal government and with pertinent state governement agencies), brokerage fees, and printing costs of the offering memorandum and various promotional materials. They are non-*amortizable capital expenditures*.

TAKE-OR-PAY CONTRACT - A (long-term) contract between a gas producer and a gas purchaser, such as a pipeline transmission company. The purchaser agrees to purchase (to 'take') a minimum annual amount of gas from the producer, or to 'pay' the producer for the minimum amount, even if the gas is not 'taken'.

In contrast, a market–out clause releases the purchaser from his obligation to 'take' gas, if there is no market for it.

TAR SAND - A sandstone in which the spaces between grains are filled with a highly *viscous* tar. [It is generally assumed that a less *viscous*, flowable oil or oil/gas mixture originally *migrated* into the sand; the gas and the lighter, more volatile hydrocarbons subsequently escaped (into the atmosphere?); the residual tar is what was left behind.]

TAX PREFERENCE ITEMS - Certain items of income, or special *deductions* from *gross income* which are given favored treatment under Federal tax law. If the use of tax preference items reduces a taxpayer's taxable income below specified levels, this may give rise to *alternative minimum tax.*

TAX LIABILITY - See *taxes due.*

TAXES DUE (TAXES PAYABLE):

Gross Income	
- Deductions	
Taxable Income x Tax Rate =	Tax Liability
	- Tax Credits
	Taxes Due (Payable)

TAXABLE INCOME - See *taxes due.*

TERTIARY RECOVERY - See *secondary recovery.*

"THIRD FOR A QUARTER" - Sometimes also known as a'quarter for a third'. A widely used arrangement for promoting an oil deal to another party. Involves *carried interests*, *working interests.* An adequate explanation of the many variations is beyond the scope of this book. *

Please refer to MONEY IN THE GROUND for complete discussion and examples of working interests.

TIGHT ROCK ('TITE') - Potential *reservoir* rock whose *porosity* and *permeability* turn out to be insufficient for commercial production of oil or gas.

TIME VALUE OF MONEY - The concept that a dollar in hand today is worth more than a dollar that will be received in some future year.

TITLE - The combination of factors that, together, constitute legal ownership of a property.

TOOL PUSHER - The supervisor of drilling rig operations. Also called 'drilling foreman' or 'rig superintendent'.

TOP LEASE - A (conditional) type of *lease* that may be granted by the *mineral-rights* owner of a property while a pre-existing recorded *lease* of that property is nearing expiration, but nonetheless is still in effect. The top lease would become effective only if and when the existing *lease* expires (or is terminated).

TOTAL DEPTH (TD) - The maximum depth of a *borehole*. Before the well has been drilled to its planned total depth, the 'current TD' changes from day to day as the well is drilled deeper. If the deeper section of a well is *plugged* off to facilitate mechanical operations on the shallower section, the top of the cement *plug* would be the '*plugged* back *TD*' (as distinguished from the drilled *TD*).

TOWNSHIP - A square tract of land six miles on a side, it consists of 36 *sections* of one square mile each. *

TRANSFER RULE - When an interest in an oil and gas property already proven to be capable of commercial production is transferred, the transferee taxpayer is generally not entitled to percentage *depletion*, although he may still be entitled to cost *depletion*, in computing his *depletion allowance* deduction from *gross income*.

TRAP - A natural configuration of layers of rock where non-porous or impermeable rocks act as a barrier, blocking the natural upward flow of buoyant *hydrocarbons* from underlying *reservoir* rocks. Most oil and gas fields are trapped accumulations of oil and gas. *

TRIP - Making a 'trip' is the procedure of pulling the entire string of drill *pipe* out of the *borehole* and then running the entire length of *drill pipe* back in the hole. This is done to change *drill bits*, to prepare for *coring* operations, etc. As the *drill pipe* is pulled from the hole, the *drill bit* acts like a piston, sucking fluids out of the *formations* and the *borehole.* Gas which is swabbed into the borehole in this manner is called 'trip gas'. When drilling is resumed, and the *mud* is *circulated* up out of the borehole, an inexperienced *mud logger* might inadvertently record 'trip gas' as a true show of gas.

TUBING - Small diameter pipe, threaded at both ends, that is lowered into a completed well. Oil and gas are produced through a string of tubing (which can be periodically removed for maintenance).

TURNKEY - A drilling contract that calls for a drilling contractor to drill a well, for a fixed price, to a specified depth and to adequately equip it so that the *operator* need only turn a valve and oil will flow into the tanks (like when a building contractor builds a house and turns over the key to the purchaser when it is completed).

The purpose of drilling a well by turnkey contract may be related to the timing of Federal income tax *deductions*. For income tax purposes, *expenses* are *deductible* from *gross income* as they are incurred. When a turnkey contract is entered into toward the end of the current tax year, the drilling costs may be pre-paid at that time. The idea is to give a *working-interest* owner (or investor) in the well, the opportunity to *deduct* the intangible drilling costs (of the well to be physically drilled during the next tax year) from his *gross income* in the current tax year.

UNASSOCIATED GAS - *Natural gas* from a gas *reservoir.* See *associated gas.*

UNDERWRITER - One who guarantees the sale of *securities* to investors. He is at risk to the extent he assumes the responsibility of paying the net purchase price to the seller at a pre-determined price. He charges a fee or this service.

UNDIVIDED INTEREST - Assume AL owns a 100% working interest in a *section* (640 *acres*) and is willing to sell a 25% (= 1/4) interest to BOB.

DIVIDED interest: AL sells a 100% interest in the northwest quarter the *section* (a 160-*acre* tract) to BOB but keeps for himself, 100% of the interest in the remaining three quarters of the section (480 *acres*). BOB owns a 25% divided working interest in the section, and AL owns a 75% divided interest in the section. (A *section* is 640 *acres.*)

UNDIVIDED interest: AL sells to BOB a

25% interest in (every *acre* in) the *section.* In this case BOB has a 25% undivided interest, and AL has a 75% undivided interest in (every *acre* in) the section.

VISCOSITY - A fluid's resistance to flowing. Generally, oil is more viscous than a mixture of oil-and-gas; depending on *oil gravity*, oil may or may not be more viscous than water.

WASTING ASSETS - *Assets* that will eventually lose their value:

- though exhaustion as they are produced (natural resources such as oil, gas, minerals, and timber), or
- through the passage of time (*leased mineral rights*, patents).

WATER DRIVE - The most efficient driving mechanism to force oil and gas out of the *reservoir*. Due to natural pressures in the *reservoir*, water moves into the base of the *pay zone*, displacing the oil-water contact upward and flushing the oil ahead of it, as oil is produced. Toward the end of the life of a water drive well, the fluid that is produced, contains increasing percentages of water, until the well becomes watered out and is abandoned. See also *gas drive.* *

WEEVIL - An unglamorous adjective (or

noun) used to describe a "green" hand — anyone new and uninitiated, especially to the mechanical operations of an oil rig. As in: "That weevil geologist over there wouldn't know a Stillson wrench from a tennis racket".

Originally the term 'boll weevil' referred to farm workers who were out of a job picking cotton when insects (boll weevils) destroyed the crop. During boom times in the oil-fields, these workers came to the rigs seeking better-paying work. Although physically capable, they were inexperienced, and had no knowledge of oil-field equipment or the tools, or what they were used for.

Today, if a newly-hired worker were to lose a finger in a piece of machinery, it might be derisively said that he'd gone and done "one of them boll weevil stunts". An uncomplicated piece of equipment is still often referred to as a 'boll weevil device' or simply a 'boll weevil' (meaning that it is so simple, anyone could operate it).

WELLBORE - Physically, wellbore refers to a *borehole* (equipped for, or intended to be equipped for, production); in other words a *completed* well. In some states (notably Oklahoma) a *working interest* owner's rights and obligations might be described as (1) encompassing all oil and gas exploration and development activities within a *spacing unit*, or (2) being limited to exploration and production only from a particular wellbore (*borehole*).

WEST TEXAS INTERMEDIATE -Refers to a grade of crude oil produced in the Permian and Midland basin areas of west Texas (average parameters are 40° *API*, and 0.4% sulfur by weight; ranges are 34° *API* - 45° *API* and up to 0.5% sulfur).

The price paid for crude oil varies according to quality. In the U.S., prices quoted generally specify grade: Alaska's North Slope crude (27° *API*), California's Kern River crude (13° *API*), Wyoming *sweet*, Oklahoma *sweet*, Gulf Coast *sweet*, Michigan *sour*, Kansas *sweet*, Illinois basin *sweet*, West Texas *sour*, etc.

Overseas crudes include: Saudi Arabian light (34° *API*), Kuwait blend (31° *API*), North Sea Brent (38° *API*), Indonesia's Minas (34° *API*), Nigeria's Bonny light (36° *API*), Mexico's Isthmus (33° *API*), along with many others. See *API*, *sour crude*.

WET - A *reservoir* rock is said to be 'wet' when it contains water but no *hydrocarbons*. (Ironically, a well that encounters only wet *reservoirs*, is called a dry hole.)

WET GAS - *Natural gas* containing liquid *hydrocarbons* - commonly *condensate.*

WHIPSTOCK - A steel blocking device placed (in the bottom of) a *borehole*. As drilling is resumed, the whipstock forces the *drill bit* (as it drills ahead) to veer off at a slight angle from the more-or-less vertical *borehole* that has been drilled down to that point. The new deviated portion of the *borehole* is called a *side track*. See *directional drilling*.

WILDCAT - An exploration well drilled to a reservoir, from which no oil or gas has previously been produced in the nearby surrounding area. When the well is located far away from all previous drilling attempts, it might be called a 'rank wildcat'. These wells naturally involve a high degree of *risk*, but a small percentage of them are successful. When a rank wildcat well comes in a discovery the return on investment can be very attractive, indeed. *

WINDFALL PROFITS TAX - A U.S. Federal revenue-generating tax on the production of most domestic (U.S.) *crude oil* (but not on gas). The intent of the tax was to prevent domestic oil producers from realizing windfall profits when domestic oil prices were de-regulated in 1980, and allowed to fluctuate according to the international oil price. Before 1980, domestic U.S. oil prices were regulated by the Federal government.

The tax was repealed as part of the Trade Bill signed by President Reagan, August 22, 1988.

WIRELINE - See *logging*.

WORKING INTEREST - An interest created by the execution of an oil and gas *lease*. The owner of a 100% working interest has the exclusive right to explore for oil and gas on a tract of land, along with the obligation to pay 100% of the costs of *drilling*, *completion*, and producing any oil or gas found. The working-interest owner is entitled to all revenues from production attributable to a *lease*, after deducting *royalty interests* (and any other burdens on the *lease*). He may reduce his share of revenues by *carving out* revenue interests and transferring them to others: an *override* to an employee, for example. *

Please refer to MONEY IN THE GROUND for complete discussion and examples of working interests.

WRITE-OFF - In common usage: a reduction in *taxable income* that results when allowable *deductions* are subtracted from *gross income.*

WTI - See *West Texas Intermediate.*

ZONE - A layer of rock penetrated by a borehole that has characteristics which distinguish it from other nearby rock; '*pay one*', '*lost-circulation* zone', 'high-pressure zone', etc.

Italicized terms are defined elsewhere in this book.

* Please refer to *MONEY IN THE GROUND* for illustrations, diagrams, and examples of any item indicated by an asterisk (*).

You will also find ... References, Sources, Recommended Reading ... along with a wealth of other practical information in *MONEY IN THE GROUND.*

See special discount offer on page 118.

Additional Information

PETROLEUM TERMS

LIQUID PETROLEUM

Crude Oil is liquid petroleum

Sour Crude . . . . contains sulfur or sulfur compounds

Sweet Crude . . . . is free of sulfur

Tar . . . . is heavy, viscous crude oil

GASEOUS PETROLEUM

Natural Gas is gaseous petroleum

In the reservoir:

Wet Gas . . . . . . contains liquid hydrocarbons

Associated Gas . . . . . . occurs naturally along with crude oil

Solution Gas . . . . is dissolved in crude oil

Gas Cap . . . . . an accumulation of free gas, above an oil accumulation already saturated with solution gas

Dry Gas . . . is devoid of liquid hydrocarbons

Unassociated Gas . . . . occurs naturally without crude oil present

At the Surface:

Casinghead Gas . . . is gas separated from the oil produced from an oil well

Dry Gas . . . . . . gas from which liquids have been separated

Sour Gas . . . . contains sulfur or sulfur compounds

Sweet Gas . . . . is free of sulfur

CONDENSATE is a gas in the reservoir.

Due to changes in pressure and temperatures, it becomes a liquid at the surface. Also called 'natural gas condensate'.

In the United States, amounts of oil, gas, and condensate are normally measured by volume, at surface temperature and pressure.

Oil and condensate are measured in barrels (42 U.S. gallons), sometimes referred to as stock tank barrels (STB). Gas is typically measured in thousands of cubic feet (MCF).

In certain overseas areas, oil, gas, and condensate are measured either by volume or by weight (according to the metric system):

1 Barrel	=	159 Liters
1 MCF	=	28.32 Cubic Meters

1 Metric Ton = 6.297 Bbls (of water @ 60° F)

In the case of oil, conversion depends on the density (API gravity) of the oil:

25° API = 6.96 barrels / metric ton

40° API = 7.62 barrels / metric ton.

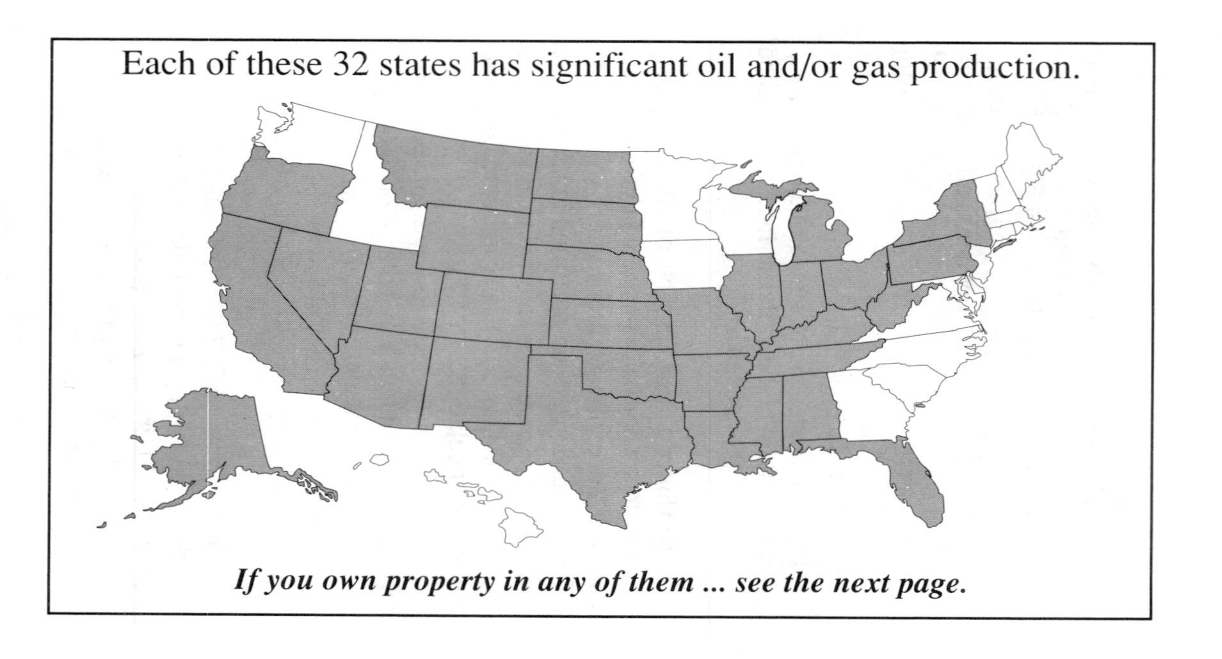
Each of these 32 states has significant oil and/or gas production.
If you own property in any of them ... see the next page.

This table is provided to illustrate the economic significance of owning mineral rights in areas with significant oil and gas production.

WELL Name	LETA GLASSCOCK # 10		ZIPPERRER #1		COTTONWOOD CREEK #1	
Date / Location	May-90; Zavala Co, TX		Jul-88; Pittsburg Co, OK		Nov-88; Carter Co, OK	
Operator	Winn Exploration Co, Tx		Amoco		CNG Producing	
Depth to Production (in feet)	6,700	ft	5,000 -12,000	ft	8,500	ft
Assumed Spacing Unit (in Acres)	**320**	**acres**	**640**	**acres**	**160**	**acres**
GROSS PRODUCTION	——	**OIL**	——	**GAS**	——	**OIL**
Reported Maximum Daily Potential *1	5,472	Bbls per Day	72,000	MCF per Day	3,700	Bbls per Day
Assumed Value of Product:	$25.00	/ bbl of oil	$1.50	/ MCF of gas	$25.00	/ bbl of oil
Projected Value of Max Daily Prod'n	**$136,800**	**per Day**	**$108,000**	**per Day**	**$92,500**	**per Day**
LANDOWNER'S ROYALTY (Assumed) *2	**1/8**		**1/8**		**1/8**	
Landowner's Share of Revenue in:						
$ / DAY for 1 ACRE	$53	/Acre/Day	$21	/Acre/Day	$72	/Acre/Day
$ / YEAR for 1 ACRE	**$19,505**	**/Acre/Year**	**$7,699**	**/Acre/Year**	**$26,377**	**/Acre/Year**
$ / DAY for the Spacing Unit	$17,100	/Day	$13,500	/Day	$11,563	/Day
$ / YEAR for the Spacing Unit	**$6,241,500**	**/Year**	**$4,927,500**	**/Year**	**$4,220,313**	**/Year**

* 1 Maximum Daily Potential — Reported Initial Potential Tests / Calculated Maximum Open Flow Potential. Actual sustained production would be less, and in any case is subject to allowable production rates set by state regulating agencies.

* 2 Assumed Fractional Interest — Although a landowner's royalty of 1/8 is assumed here, royalties of as much as 3/16 or more are not uncommon.

** Drilling Technique — The Leta Glasscock well was completed with a horizontal extension of about 2,625 feet.

*** *For Illustration Purposes* — These are examples of some of the best wells drilled in recent years onshore in the U.S.

MONEY IN THE GROUND

Is the Critic's Choice . . . .

"...don't lay a nickel down on the table until you've read this book."

DOUG BENTIN – Columnist, Oklahoma Gazette

"... contains some of the clearest and best-written explanations of oil and gas investments we have ever seen ..."

LIMITED PARTNERSHIP INVESTMENT REVIEW
Springfield, New Jersey

"... designed for the potential investor. Helpful and simplified ... adequate treatment of every relevant aspect from geology to taxation."

PETROLEUM ECONOMIST – London

" .. new and comprehensive. Provides specific deals with examples."

OIL & GAS JOURNAL – Tulsa, Oklahoma

" .. If you're an investor thinking about gas and oil ... you must get MONEY IN THE GROUND ...as comprehensive a book on the subject as you will ever find."

ALAN CARUBA – Columnist, Essex Journal

Listen to what the experts are saying....

"MONEY IN THE GROUND ... is now the book I would heartily recommend to anyone interested in learning about oil and gas investment."

ROBERT A. STANGER – Chairman of the Board
Robert A. Stanger & Co., Shrewsbury, NJ

"I was looking for just this sort of book ... both up to date and written in a style which a 'non-oily' can understand ..."

GENE A. CASTLEBERRY – Partner
Castleberry & Kivel
Attorneys at Law, Oklahoma City, Oklahoma

"... we rate your book as excellent ... and would recommend it to others."

WILLIAM R. McHUGH – President
Royalty Information Systems, Inc.,
Covington Louisiana

"Excellent value for money. An excellent informative book."

JOHN L. PIESEK –Vice President
Stockmen's Bank and Trust Co.,
Gillette, Wyoming

Always patronize the establishment where you purchased this volume — or your local bookseller — first, for any title.

However, if a MERIDIAN PRESS publication is temporarily out of stock, you can order direct, using this easy form:

HERE'S HOW TO ORDER

Let's talk an OIL DEAL!

Your Key To Oil Patch Lingo

$30 for three copies *

[Single copies: **$13.50** each = $10.00 plus $3.50 shipping and handling.]

MONEY IN THE GROUND

Insider's Guide To Oil Deals 3rd Edition

$35 for each copy *

* **Prices in bold** are U.S. currency, and include shipping and handling in U.S. [Overseas orders please add $10.00.]

Send to:

MERIDIAN PRESS - LT
PO Box 21567
Oklahoma City, OK 73156
USA

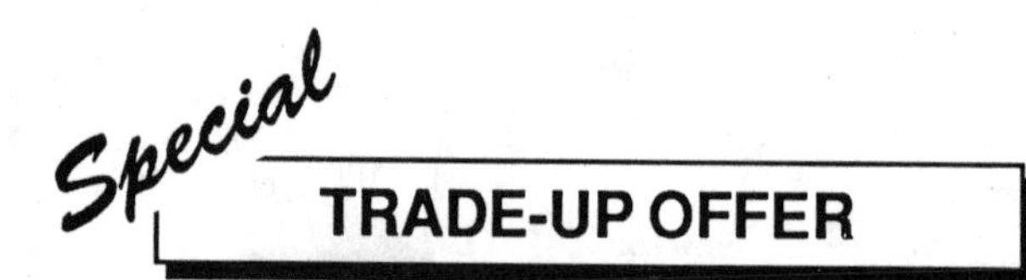

The book you are now holding contains just a fraction of the *valuable information* and *practical tips* you will find in:

MONEY IN THE GROUND
Insider's Guide to Oil Deals.
3RD EDITION $35.00

Trade-up . . . and *SAVE $5.00!*

To get your copy of the *current edition* of ***MONEY IN THE GROUND*** at **$5.00 off** the regular mail order price:

Place a copy of this special offer ... along with your check for **$30** ... in an envelope, and mail to:

MERIDIAN PRESS
Preferred Customers
PO Box 21567
Oklahoma City, OK 73156

Order your copy, *today!*

Contents

About the author . . .

John Orban, III was born in Denver. He graduated from Princeton University, having worked summers as a roustabout in Morgan City, La., while majoring in Geological Engineering. After serving in Vietnam, he worked overseas as a mud-logger.

Mr. Orban was International Explorationist with a major oil company in Dallas, and then in London, before joining a start-up exploration oil company in Oklahoma City.

Active in the U.S. oil business the past ten years, Mr. Orban is author of the popular *MONEY IN THE GROUND*. As Petroleum Consultant, he specializes in oil and gas investment. He is married and lives with his wife in Oklahoma City.